Rescue Pilot

The Daring Adventures Of A New Zealand Search And Rescue Pilot

John Funnell

16

EasyRead Large

RHYW

Copyright Page from the Original Book

RANDOM HOUSE

UK | USA | Canada | Ireland | Australia
India | New Zealand | South Africa | China

Random House is an imprint of the Penguin Random House group of companies, whose addresses can be found at global.penguinrandomhouse.com.

Penguin
Random House
New Zealand

First published by Penguin Random House New Zealand, 2016

10 9 8 7 6 5 4 3 2 1

Cover and text design by Kate Barraclough © Penguin Random House New Zealand
Prepress by Image Centre Group

Printed and bound in Australia by Griffin Press, an Accredited ISO AS/NZS 14001 Environmental Management Systems Printer

A catalogue record for this book is available from the National Library of New Zealand.

ISBN 978-1-77553-849-3
eISBN 978-0-14377-006-0

MIX
Paper from responsible sources
FSC® C009448

TABLE OF CONTENTS

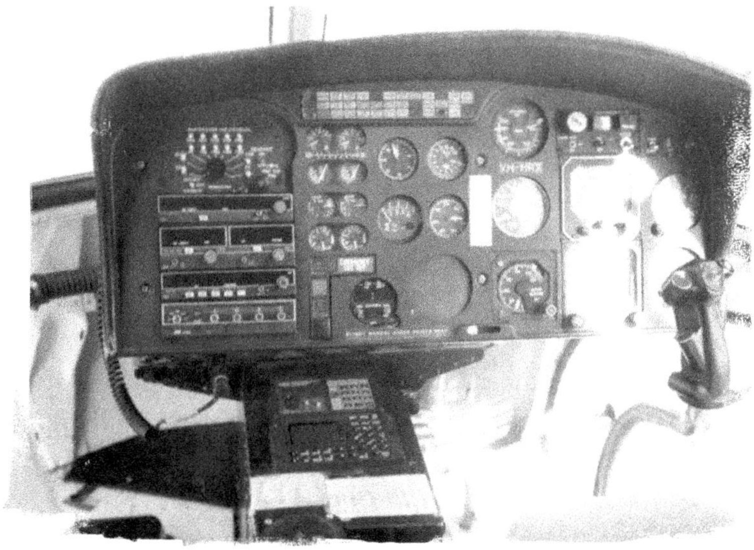

PROLOGUE

Somewhere, up ahead, someone is bleeding, but I have to put that out of my mind. My job is negotiating with time and space. I have my clock, they have theirs. And, in the end, whether the rate at which my clock is ticking matches his is out of my control. I'm just the taxi driver here.

There are factors that are within my control. My eyes scan the instruments from time to time as reflexively as my eyelids lower to blink. Condition normal, they tell me. Revolutions per minute are normal, if toward the higher end of the range. Airspeed is satisfactory, and while I can only guess at speed over the ground until I regain GPS coverage, I assume I'm making good. Oil pressure, temperature, normal. Fuel burn ... well, Grant Biel is sitting beside me doing the calcs, and that is a big comfort. He has an almost instinctive feel for the rate at which the engine drinks jet fuel. He's talking to Dave Brown in the Piper Navajo we have overhead, which is monitoring wind direction and speed, looking for anything that might blow in our favour.

Every now and then my mind strays to consider—another reflex—what would happen in the event of engine failure, or some other such disaster. The answer is pretty simple, really:

I'd auto-rotate into the Southern Ocean, somewhere in the darkness down there below, and hope to arrive in good enough shape to launch the raft and get clear of the aircraft. But I don't dwell on this. I've practised the auto-rotation enough times, and I've mentally rehearsed the other steps time and again over the years. And I have implicit faith in the maintenance team and their systems. Of all the things that can go wrong, mechanical failure is the least likely.

I think ahead. That's where the real danger lies.

If I were the worrying kind, I'd be wondering whether there would be cloud cover, and whether I'd find a way down. I'd be thinking about the wind velocity and obstructions in the vicinity of the landing zone. I'd be fretting over the quality of the landing zone itself, and whether I'd be able to set down close enough to the injured man to be useful.

But I'm not worried. There is no point in worrying about any of that, as it's out of my control.

The thoughts that are hardest to put out of my mind are those of the man himself. His injuries are massive, I've been told. He's losing blood and, as usual in these emergencies, the clock is ticking. There have been lots of studies showing how vital it is to get timely assistance to trauma patients. Paramedics speak of 'the golden hour', by which they mean the dramatically improved outcome when advanced medical care reaches the afflicted person within 60 minutes. For every minute after that hour has lapsed, the chance of a positive outcome falls away.

I can't help doing the maths: distance to destination divided by likely speed over the ground yields the remaining flight time. For the man I'm beating towards, the golden hour lapsed long ago, and many more minutes will pass before I reach him.

As usual, there's a little surge of adrenaline at the thought. But it's important to keep focused on the here and the now. He has his clock, I have mine.

There are certain stock phrases that I use to describe to the press what must be done: keep the aircraft right way up and roughly level. And when people tell me that my job must be tremendously exciting, I tend to reply that it's a mixture. Hours of boredom, I tell them, and moments of terror.

Grant has word from Dave that there is a favourable wind at higher altitude, so I put the helicopter into a steady climb. Other than that, there is nothing much more to do than scan the instruments and keep the helicopter pointed toward the patient. There's no point in wasting sympathy on poor Pat Wynne, who's crammed in back there amongst his medical supplies, his kit and the three drums of fuel. He's been with me on so many of these jobs, he knows the drill.

I look ahead. There's nothing to see, of course, because the sun isn't due to rise for another three hours, but it's habit. And it's because I know that somewhere in that direction a man is bleeding.

My first venture into rotary wing aircraft: a Bensen gyrocopter I built from plans in the 1970s.

CHAPTER ONE

AB INITIO

I was high above the ground, relatively speaking. I was in a tree at the back of the property. I reached for a branch, missed my grip and, with a terrible swooping feeling in the pit of my stomach, I fell.

My first landing wasn't one of my better ones. I struck a picnic table on the way down, my right arm taking the full impact. The air was driven from my lungs. I lay there stunned for a minute or two, as my brain tried to catch up with what had just happened to my body. Tough, tree-climbing kids don't cry. After I told Mum what happened, she took me to Palmerston North Hospital, where my arm—broken, as it happened—was seen to. The doctor heard about the tree. He heard about the fall.

'What were you trying to do, young fellow?' he asked, not unkindly. 'Fly?'

I nodded sheepishly.

***

I grew up on my parents' dairy farm at Oroua Downs in the Manawatu, about 10 kilometres north of Foxton. The Funnells came from Australia most immediately: my granddad and grandmother came over on the *Monowai* in 1905, arriving in Wellington. They settled at Raumai near Ashhurst, where Granddad plied his trade as a blacksmith building bridges and shoeing horses. My grandmother, meanwhile, steadily built up a herd of dairy cows, and it was largely the proceeds of her milking operation that enabled them to put a deposit on the 245-acre property at Oroua Downs in 1919. So that's where my dad, Lionel, and his two brothers, Monowai and Darcey, grew up, too.

Probably influenced by his own father's practical bent, Dad learned how to use a forge. (I can recall as a young boy having to wind furiously on the handle that drove the fan, forcing air into the forge to heat the metal that Dad was working.) When World War II broke out, he enlisted in the army and was assigned to the 6th Field Company of the New Zealand Expeditionary Force, holding the rank of sapper. His particular area of expertise was in clearing minefields.

Late on the night of 1 August 1944, he and his unit were sweeping for mines ahead of the Allied infantry advance on a ridge south of Florence. They came under sustained fire from a building in their path, and Dad volunteered to move forward and sweep the courtyard for mines so that the Allied tanks could mount an attack on the position. He performed this task 'calmly ... with courage and with complete disregard for his own safety', according to the citation I saw many years later. He won the Military Medal for that episode. As a kid, I gathered he had done something pretty special from the fact that there was a large photograph of him hanging in the Oroua Downs' hall. But it wasn't until an old bloke knocked on my door a few years ago and asked if I was Lionel Funnell's son that I learned the details. Reg Burrows had enlisted with Dad and while not there with him that night in Italy, he knew well what Dad had done and gave me a copy of the citation.

Typical of war veterans, Dad never talked about his experiences in the war. Over the years, he and I spent a lot of time together milking cows, and the topics of our conversation ranged far and wide. He talked about the war in general—I was fascinated by how Hitler managed to get ordinary people to do the things they did—but he never talked about his own experiences. They weighed upon him. As kids, we soon got to recognise when he needed space, which he did from time to time. Mum told us not to ask him about his war, and we didn't. Most holidays, we would load up our old Chev car and head south, stopping here and there so that Dad could catch up with old army mates. I suppose it was hugely relaxing simply to be with people who knew, without having to be told, what he had gone through.

My mother, Beth McGregor, was born in Blackball on the West Coast of the South Island. Her dad was a gold prospector and a carrier, using a truck with solid rubber tyres. During the war, he added another string to his bow and became a taxi driver—so that he could get an extra ration of petrol. Mum trained as a schoolteacher and was posted to Foxton, where she boarded with a baker, Johnny Johnson, and his family. Johnson had a little caravan that he used to take around the shows and other local events, selling his wares. Mum was somehow talked into working a few weekend shifts in this. That's where Dad met her, after his return from the war, and that's how, when asked where he met his wife, he came to be able to tell people truthfully: 'I met her at the pie cart.'

Soon after their marriage, in 1950, I was born, the first of five. There was about a year between each of us, apart from Keith, the youngest, who took a little longer to arrive. Dad took over the farm from Granddad in 1953 and that's where I grew up.

There wasn't much in the way of entertainment when I was a kid—no TV, for example—so even though he had been up early and had worked a long, hard day, Dad used to read to us each evening before bed. While I was at primary school, we became the second household in the district to get a TV. It was black and white, of course, and there was only one channel, and that was only on air for a few hours each evening, but we were the envy of all our neighbours, our mates, and quite a few of their parents besides. We had visitors most nights.

You couldn't ask for a better childhood than mine. Whatever Dad's troubles from the war, and whatever strains they might have placed on him, Dad and Mum were devoted parents, kind and supportive always. In fact, it wasn't really until I was well into my teenage years that I realised not everyone is as fortunate in their family background. And it wasn't until the 1980s, when I was doing hospital transfers and happened to see the movie *Once Were Warriors,* that I realised for some kids, their parents are the enemy. Nothing could be further from my own experience.

Like his own father, Dad was a practical man. When I was still at primary school, he bought a massive arc welding plant. He devised an arrangement whereby the PTO (power take-off) shaft of our Fordson Major tractor coupled to a very long, exposed shaft, at the end of which a very large, exposed pulley drove three exposed and flapping V-belts which, in turn, powered the welder. Using this, he built stuff like farm gates and yards, but he also built slides, swings, see-saws and merry-go-rounds for us. Our playground was the envy of the neighbourhood. He bought and maintained bikes for us all. I used to ride the mile or so from home to school every day, rain or shine and (usually) wind. In the morning when I set off, it was usually into the bitter katabatic that slumped off the Tararua Ranges and howled across the Manawatu. In the afternoon, I was usually pedalling into an even stiffer westerly sea breeze—and all this on a fixed-gear bike.

***

I wasn't exactly a star pupil at Oroua Downs School. There were 25–30 of us in the class and I was pretty much bottom of the heap in everything except maths, for which I had a faint aptitude. I just seemed much slower to catch on than everyone else, particularly the girls. Some of the girls would polish off a piece of work in a third of the time it took me to do it. One day, I think it was in Standard Four (Year 6 in today's money), one of my teachers explained it all for me.

'Funnell,' he said. 'The trouble with you is you're just dumb.'

Well, that was a load off my mind! If I was just dumb, there was no need to bust a gut to try to achieve beyond the natural scope of my abilities. I enjoyed helping Dad on the farm: from now on, I vowed, I would bide my time at school, eating my lunch and playing sport, until I was old enough to go farming. After all (or so I thought), you didn't have to be bright to be a farmer.

***

One day, while in my very early school years, I was attracted by the spectacle of a Tiger Moth biplane spraying our neighbour's lupins. The Nicolsons had a large area covered in these bushes that prevented grass from growing, and they had marked out their boundaries with sheets of newspapers to guide Frank Desborough spraying the lupins with a mixture of diesel and water. I watched him perform a turn, sweep in and dump a dirty-looking plume of spray on the far side of the paddock, and then bank and line up for another run. I watched as he did it all over again. Like most of the pioneering top-dressers, Frank had learned to fly when the war broke out, and served in the air force.

It didn't occur to me, watching him soaring and swooping over the Manawatu that morning, that pretty much anyone could learn to do what he was doing.

Years later, while biking to school, I heard and observed a Cessna Agwagon taking off from the airstrip near the road to school. I didn't wear a watch in those days, and it was a bit of a shock when I eventually got to school and found I was over an hour late. But my explanation must have impressed my teacher, because she took the whole class along the road to watch the plane working.

It was probably around then that I began to daydream about flying.

***

I don't know whether it was because they saw some sort of potential in me, but Mum and Dad made a lot of sacrifices for me. It would have been much cheaper just to send me to a local high school, but instead they paid significantly more for me to attend Feilding Agricultural High School. I remember Dad writing out a cheque with his big, calloused hands sitting at his writing desk. I wasn't thrilled about being sent away from home (Feilding Ag was a boarding school), and I flatly refused to go into the school dorms. So for my first year there—third form—a room was found for me in a private home up the road from the school.

In fourth form, I became a boarder. This was an eye-opener for someone who had known pretty much nothing other than love and affection. On one of my first days boarding, a big, solid fifth-former by the name of Edwards marched up to me and told me to polish his shoes. I was much smaller than him—I was much smaller than most kids there—but I shook my head.

'I don't polish anyone's shoes,' I said.

Predictably, he whacked me.

Less predictably, I whacked him back, good and hard in the guts. He wasn't expecting that, and the surprise registered on his face, a few seconds before he scowled and set about giving me a hell of a hiding. But it was worth it in the long run. Word went around: watch out for that little fucker Funnell! I got left alone, and, in fact, I traded on my reputation as a bit of a firebrand to tell the bullies—there were plenty of them—to leave the rest of us smaller chaps alone.

Besides the usual curriculum subjects—maths, English, social studies and the like—Feilding Ag taught things like animal husbandry, pasture management and other subjects that would be useful to someone setting out on a career in agriculture. Better yet, they owned real working farms—a beef and a dairy unit—and you could get out of the classroom by volunteering on the farms. I wasn't in the top class and I wasn't top of the class I was in, so that's what I did.

Around the end of my fifth form, Dad developed a gallbladder infection. It's no big deal these days, and the operation is only a few times more debilitating than getting a tooth filled. But back then it was a pretty major operation, and Dad took a few weeks to recuperate. I took time off school to run the farm while he was laid up, and by the time he was back on deck, the school year had finished.

Without anyone sitting down and actually talking about it, all parties more or less agreed that my schooling was over, and that my career in agriculture had begun. I was sixteen.

We ran about 120 cows, and the property was on very light soil. They would irrigate it these days, but before irrigation became commonplace, blocks like that tended to dry off pretty badly in the summer. Ours was often pretty marginal for dairying purposes. All the same, it wasn't a bad life, and I enjoyed my time in the shed with Dad. The trouble was, I was young, and while the sound of the milk chugging quietly along the lines into the vat was music to Dad's ears—money in the bank—it only got in the way of my daydreaming.

By now, I was definitely dreaming of flying.

***

One of my extra-curricular activities at school was Boy Scouts. I had become a Venturer Scout and I was fortunate enough to have a scoutmaster by the name of Colin Alexander. One day, Colin took me aside.

'Thought you might like to know, John. I've heard that Venturer Scouts are about to set up an annual summer flying school at Matamata.'

He must have read in my face the impact this news had on me.

'Applications have closed,' he said. 'They're full up. But leave it to me.'

Colin was an energetic individual, a cheerful, whistling builder by day but a demonically effective lobbyist and a tireless letter-writer by night. He brought all his energy, expertise and sheer persistence to bear in getting me a place on the flying programme.

'Dear Mr Alexander,' the early replies from Scouts HQ in Wellington read. 'Thank you for your recent letter regarding the application of John Funnell to join our flying school. We regret to advise...'

Not to be discouraged, Colin would fire off yet another letter pleading my cause. Wellington soon realised they were in a full-scale war of attrition, and finally, one day, a letter arrived in Colin's mailbox.

'Dear Pest,' it began, and went on to offer up unconditional surrender. I was on the flying programme for the first ever Walsh Brothers Memorial Scout Flying School at Matamata.

I was thrilled. Mum, who only ever seemed to notice the news reports of planes falling from the sky and crashing and burning around the country, didn't share my enthusiasm for aviation. She was against my going. But with Dad behind me, I was all clear for take-off.

***

My first flying instructor was a World War II RAF veteran named Gordon Reader. Gordon had flown Harvards and the Lockheed T33 jet fighters in the latter part of the war. He had come out to New Zealand shortly before I signed up for the flying school, and while working at Palmerston North as an air traffic controller, he moonlighted as an instructor. He became a bit of an identity in New Zealand aviation circles, and in later years owned three vintage aircraft, which he named 'Rugby', 'Racing' and 'Beer'. He was a hell of a nice bloke.

Gordon taught me and one other in a Champion 7-EC, a purpose-built trainer, a little two-seater, high-wing affair belonging to the Middle Districts Aero Club. There were too many aircraft for everyone to fit in the circuit at Matamata. Because the Champ was a tail-dragger—it had a wheel at the tail instead of at the nose—it was suitable for use on shorter airstrips, and we used to go from Matamata to Te Aroha for our practical lessons. The first time I went up was incredibly exciting. The only other time I had flown was when I travelled in a National Airways Corporation DC3 to Christchurch to visit my grandmother, which was very different to the small, noisy Champ. Because the seating arrangement was tandem and the student sat in front, I felt as though I was alone, which was thrilling and would have been scary had it not been for the calm, kindly English voice behind me telling me what to do. We would 'touch and go': taxi, take off, do a circuit of the strip and then come in to land again. Then we would repeat it, getting through perhaps 10 circuits in a lesson. I couldn't get enough of it, and because I was used to getting up early to stare at cows' rears, I was always one of the first to present myself, bright-eyed and bushy-tailed, at the airstrip.

One day—24 January 1967—after we had come to rest on the runway, Gordon climbed out of the plane. 'Stay there,' he said, as I undid my belt and began to do the same. 'Take her up and do it again.'

'By myself?' I asked, incredulous.

'Yes,' he nodded. 'By yourself.'

He might have added something—a bit of advice, perhaps—but I was hardly listening. I was both elated and terrified. Yet even though I had only had around seven and a half hours' flying, I felt I knew what to do.

I taxied out, opened the throttle, brought the Champ to speed and then eased back on the stick. At that moment, as I shot into the sky at an alarmingly steep angle, I realised what Gordon's final piece of advice had been. Since his weight was no longer in the seat behind mine, I needed to trim her forward a bit.

I made the necessary adjustments, performed the circuit and then managed to land her again, with only a wobble left and a wobble right. That was it! My first solo flight, and I was at the legal minimum age of 16!

I did a few more consolidation flights before my time at the aviation school was over. My sights were now firmly fixed on my private pilot licence (PPL), and that meant getting more airtime. Dad was great about this: he let me take his big old Ford Falcon and fill it up on his account at the local Himatangi service station so that I could drive to the aero club aerodrome at Palmerston North, where I flew with Gordon Reader, Peter Strugnell, Dick Basset, Errol Smart and others.

You needed 50 hours' flying time to qualify for a licence, and that wasn't cheap. Dad was paying me £9 per week, but it cost £6 a week to fly. There wasn't a lot left over, so while my mates were all taking girls out, I was stuck penniless at home. In a way, that was a good thing, because the other requirement for my licence was the exam. I'd never sat an exam in my life—too dumb, remember—so I knew I would have to work a hell of a lot harder than anyone else. So that's what I did. I went to night classes, and I studied my backside off.

There was only one exam for the PPL in those days, and it consisted of questions on all five of the theoretical subjects—navigation, meteorology, principles of flight, aircraft technical knowledge and performance—and while I entered the exam room with a sense of doom, I found that I knew the answers to a lot of the questions, or at least I thought I did. Still, I didn't let myself believe I'd pass until I actually got the result through the mail. I had passed, after all.

'Not bad for a dumb kid,' I told myself. 'Not bad at all.'

There was still the matter of the actual flight test. I sat for it on 28 April 1969 with Harold Bennett. My heart sank when I learned it was Harold: he was another military man, and he was known as a tough man to get past on a flight test. He showed no emotion, and you either passed or failed with him. Well, despite my nerves, I managed to pass.

A few years later, I read that Harold managed to put an aero club plane through a fence when he asked a pilot in a flight test to demonstrate a simulated engine failure into a paddock. The engine didn't respond to the command for power when on the overshoot, so the simulation became the real thing for both on board. Harold, desperate, took over the controls—and hit the fence. That incident forced a slight change of procedure for flight tests. From then on, it was made clear the student was the pilot in command during a flight test. The instructor would only take control if they thought you were going to hurt or kill them.

***

Staring at cows' backsides was less appealing than ever now that a new career seemed to be opening up in front of me, one in which the sky, quite literally, was the limit. By now I had thoughts of becoming an agricultural pilot, so I looked about for work as a loader-driver—a kind of apprentice pilot who assists an agricultural pilot by loading the plane's hoppers with whatever payload was to be spread, be it herbicide, or superphosphate, or rabbit baits.

I learned of a job going with Manawatu Aerial Topdressing: I applied to Basil Forster-Pratt, who owned the company, but he must have got in touch with Mum for a reference. Mum, who hadn't got any fonder of the idea of me flying, doubtless did a number on me, and I missed out on the job. But then I heard of another loader-driver position for Wanganui Aero Work based at Pahiatua and working the Wairarapa. Doug Bartlett, the area manager at the time, did a quick interview to confirm I had a heavy traffic licence. Then he introduced me to Lindsay Williamson, the pilot, and I had the job. I found a family to board with: the Renwicks had two sons younger than me, and were wonderful people—I fell on my feet with them. So I shifted to Pahiatua to get a toehold in the agricultural aviation industry.

The job entailed getting up before sunrise and driving an RL Bedford truck loader to the airstrip, where I would load up the Fletcher 300-horsepower aircraft for Lindsay. He would take off, deliver the payload and then return for another load. You'd do that all day, stopping only to have lunch and morning and afternoon teas, which the farmer supplied. By the end of the afternoon, the Bedford, which carried 1200 litres of avgas, got low on fuel, and I would take it back to base in the evening and fill it up ready for the next day. The Wairarapa isn't exactly an agricultural pilot's paradise: there were plenty of down days due to high winds or rain. To make sure I kept up my flying hours, I decided to buy a plane of my own so that I could make the most of my spare time. There was a Piper Cub J3 (registration number ZK-AHC)—to all intents and purposes, a very similar plane to the Champion—going cheap in Dannevirke, so I bought it. I flew ZK-AHC about the place for a while until one day when I had just landed in a strong breeze in Palmerston North. I was probably pushing my luck flying at all that day, but when you're young you make a few questionable decisions. I was taxiing in when the wind got under a wingtip and before I knew much about it, the plane was on its back. I struggled out and ruefully inspected the damage, which was bad enough to make the plane unairworthy. Luckily for me, it was insured.

There are light aircraft and then there are light aircraft: the Cub was one of the lightest of the lot, and notoriously prone to this sort of episode. A couple of years ago, I tracked ZK-AHC down and bought it back. It's in pretty good nick, but it has been rebuilt several times after performing similar flips.

Any crash you walk away from makes you a much better pilot. After that, I wasn't quite so inclined to ignore the advice that older and more experienced pilots would offer me.

***

The next step for me was a commercial pilot licence (CPL). I was well on the way to having collected the necessary 200 hours' flying time, but I still needed to polish up for a cross-country flight test with a man from the Civil Aviation Division of the Ministry of Transport. Worse, from my perspective, were the exams: this time, each of the five subjects had its own exam.

I took three months' leave from my job during the winter of 1970 and enrolled in a nine-week course at the Commercial Pilot School run by the Wanganui Aero Club. I got a room under the roof of an elderly lady in Wanganui who must have been very agreeably surprised by the character of her boarder: while every other young bugger in town and in wider New Zealand was out on the tiles most nights of the week, and especially at the weekend, I was at home, swotting. I just presumed that as a dumb guy, I needed to put in the extra hours. Every now and then I would go out for a quiet beer, but when everyone else was kicking on,

I would grin sheepishly, shake my head and go back home to my books, knowing that the next day I would be under the watchful eye of Ivan Warmington.

Ivan was yet another war veteran, and a distinguished one at that. He had flown Lancaster bombers, and was awarded the Distinguished Flying Cross for his service. He was a nononsense sort of bloke, but a real gentleman, and calm. After what he'd been through, I suppose nervous teenagers at the controls of a light aircraft weren't going to scare him.

There were trial exams you could sit to prepare for the real thing, and I seemed to be doing okay in those. My usual failure to catch on was obvious only in meteorology, but I soon had a contingency plan. I wouldn't bother studying 'met' at all, and would instead concentrate on the other four subjects. I would sit all five exams, and when (inevitably) I failed met, it would be good practice for my real tilt at it next time around. I duly sat the exams, and went back to Pahiatua to get on with my job.

One evening, Mum phoned.

'John, I think your exam results are here. There's an official-looking letter for you from Civil Aviation in Wellington. Do you want me to open it?'

'Hell, yes!'

There followed the sounds of the receiver bumping on the table and of ripping paper.

A short pause.

'Oh, yes,' Mum said. 'You passed them all.'

'What?' I said. 'All of them? All five?'

'Yes, all five.'

'Five passes?' I couldn't quite believe it. 'You're sure?'

'I think I can read, John,' Mum said, with the slightest trace of schoolteacher impatience. 'There are five subjects here, and each of them has the word "pass" written beside it.'

'What did I get for met?' I asked.

'Let's see. Meteorology. Seventy per cent.'

I was gobsmacked.

Now, that—I told myself and probably Mum as well—now, *that* wasn't bad for a dumb guy.

Spraying pine trees in the central North Island.

CHAPTER TWO

WINGING IT

With my commercial pilot licence and a few hours under my belt, I was asked to shift to the Taihape base of Wanganui Aero Work (WAW) to drive a loader truck for John Harding, who flew a de Havilland Beaver as well as a Cessna Agwagon and a Cessna 185. John mostly did top-dressing work over the sheep stations spanning huge swathes of country on the Central Plateau. Plenty of people told me to be careful. John had a reputation for being a demanding employer, but I decided I'd have a crack.

At the interview, John was no-nonsense and with a firm handshake and lots of direct eye contact. He told me that if I got the job, I could expect to work pretty hard, but that he would see to it that I got opportunities to further my flying career whenever possible if I shaped up well as a loader-driver. And so, after he offered the job and I accepted it, it proved to be. He was one of the best employers I ever had.

John flew out of an airstrip that used to belong to the pioneering top-dressing outfit Rangitikei Air Services back in the day. The strip came with an adjacent house, and I moved in there with three other young bucks: Ian Cross, Ian Johnson and Graham Brough. The two Ians were telecoms technicians. Graham flew for Rangitikei Air Services (RAS), now owned by WAW. RAS was formerly owned by Rex Garnham, a Corsair pilot in the Pacific in World War II and a hell of a nice chap, a real gentleman. His grandson Davie Fielding flew for me out of Palmerston North many years later doing rescue work, until he was killed doing a flight test for a young pilot when a Cessna 150 was involved in a mid-air collision with his Robinson helicopter.

Each morning when John Harding was working—most mornings—I would get up at about five and make my way out to where the Beaver was tied down on the side of the strip right adjacent to our house and start it up. In the winter, the plane would be coated in a half inch or so of frost. I had to remove the frozen cover before I could begin the elaborate ritual of starting the plane's radial engine—a 450-horsepower Pratt and Whitney radial, which enthusiasts call a 'round' engine. Starting one entails pulling the prop through nine blades to make sure no engine oil had seeped past the rings of the lower cylinders: if you didn't take this precaution, a hydraulic effect in the cylinders could result, bending the con-rods. You then primed the engine with a few pumps on a pump just inside the cockpit door. Then you pumped the engine 'wobbly pump' to get the fuel up to the carburettor, flicked on the master switch, adjusted the mixture lever to full rich, pumped the throttle two or three times and adjusted it to a point just above the stop, flicked another switch to crank the engine over (which produces a loud whine), and after five blade rotations, switched on both magnetos. The engine would then commence backfiring loudly and the whole plane would jerk and shudder. By then, you were anxiously watching to ensure the oil pressure was coming up, and with luck, it was in the green zone on the gauge by the time the detonations from the engine had smoothed into something like a steady, pleasant roar. Now the important bit: you let the engine oil warm up to at least 40°C. To accomplish this, I'd leave the plane idling on its chocks while I went in and had breakfast, and by the time John arrived just before dawn, the oil would be nice and warm and the screen defogged. We'd climb aboard and fly over the cold tussock land to the airstrip on whichever station we happened to be working that day, usually arriving just before the first blazing fingernail-clipping of the sun appeared over the horizon.

On the ground, I would hop out and climb into the frigid back cab of the Bedford truck. I would have already set it up with a load of fertiliser in the hopper the day previously, so there would be no mucking around loading up the plane. Once John had roared off into the dawn sky, I would pull off the lid or cover of the heap of fertiliser in the strip bin and refill the truck loader bucket, ready for John when he landed a few minutes later. Every two hours, we would refuel the plane and then he'd be back into it. Apart from a couple of short breaks for lunch and a cuppa, you'd be flat out doing this all day from dawn to dusk, whereupon I would load the hopper ready for the morning and we'd fly back to base, fuel up the plane and call it a day. You could see why John had a name for being a hard man to work for: he worked hard himself, and asked no less of his offsiders. I was up for it, so we got on really well.

***

One day, after I had been working for John for about three years, he arrived at work and told me brusquely to pack my bags. I was stunned.

'What? Are you giving me the sack?' I asked him.

'No,' he said. 'It's time you started your Ag training with Berrick.'

It turned out John had arranged for me to enter Berrick Dalcombe's Agricultural Pilot Training School down in Wanganui. I was absolutely delighted.

As Mum was constantly reminding me, the period during which I was serving my apprenticeship in agricultural aviation was a particularly dismal one in the history of the industry. Lots of pilots seemed to be getting themselves killed. Berrick Dalcombe, another veteran of the war, had seen the need to establish a school to teach the skills required of the safe and effective top-dressing pilot—flying off airstrips and handling heavily loaded aircraft at low levels where there were plenty of hazards, such as trees and wires. Berrick had a Cessna 180 with a hopper aboard, and this was the main training aircraft. He also had a second pilot named Richie de Montalk who did a lot of the training and then flew with us in a dual-control 300-horsepower Fletcher FU24, a commonly used top-dresser in those days, which was so underpowered that you'd occasionally hear pilots say that it was only the curvature of the earth that made it possible for the Fletcher to stay in the air. A couple of times, Richie and I took the Fletcher up the course of the Whanganui River so that I could learn how to spot wires and safely manoeuvre amongst the hills. The rest of the time, Berrick or Richie would take trainee pilots like me out to work on the bigger airstrips and easier jobs, keeping an eye on what you did. Between them, they must have done a good job of teaching me. I managed to stay alive while plenty of other non-Dalcombe trained top-dressers didn't.

Once I had finished my Ag training, I went back to John Harding in Taihape. I got to do a bit of top-dressing of my own in John's Cessna 185. Harding would load me while keeping an eye on me, the strip, weather, wires and other issues that could catch a young, inexperienced pilot out. In the early days of this phase of things, I also had Andy Stevenson, who worked as a pilot for John based in Taihape, keeping an eye on what I was doing. He had quite a bit of experience at the controls of a top-dressing aircraft and he tutored me in his low-key way. 'You have to have a system, Jack,' he would say, 'and you must stick with that system every time. If you get interrupted, start your system again.'

Fuel, gauges, hopper open and then closed, flaps set, oil pressure and temps okay—failing to shut the hopper doors was embarrassing, because as soon as the loader-driver dumped the load in the top of the aircraft hopper, it would fall out the bottom. Then you'd have to shovel it back into the loader bucket by hand to start again. You made sure that never happened more than once. It also demonstrated you were not using the Stevenson system correctly.

Andy had another saying: 'He who dumps today lives to dump another day.' You were better to let some or all of your load go early rather than hang on till you went through the fence at the end of the strip.

'How would you tackle this block, Jack?' he would ask, as we surveyed the next section of a job to be done.

I would tell him.

'Yeah, you could do it that way,' he'd nod. 'But it would be safer and faster if you did it like this...'

Andy, John, Berrick and Richie impressed on me that you never fly up a blind gully, and that if you were approaching a hillside, you did so at an angle rather than square-on. That way, if, as so often happened, you found you'd miscalculated speed or distance or the effects of the wind, you could execute a slight turn to get out of trouble, which was far preferable to finding out the hard way that you couldn't outclimb the terrain ahead.

Clocking up hours was no problem working in Taihape. In my first year after achieving my ag pilot rating, I flew more than 1000 productive hours. There was one job I recall where, flying off a top-dresser's dream airstrip—perched on a ridge that was bound on three sides by a broad, steep gully—in the Taihape high country, we managed to set a record applying 220 tonnes of lime with a 300-horsepower Fletcher. That's nothing compared with what modern aircraft can do, but it was quite an achievement under the circumstances. I wouldn't be surprised if that record still stands for that type of aeroplane.

***

Another of the bigger changes in my life happened around this time. I first met Patricia Evernden while she was still at high school in Dannevirke. My best mate (and eventually, best man) Jeff Winter and I used to travel to Dannevirke most Friday evenings in his Morris 8 to check out the supposedly Danish girls in Dannevirke. Raspberry nip was the drink of choice in those days. Trish and I chatted a few times before I summoned the courage to ask her out. I was working long hours, but where there's a will, there's a way. We pretty quickly became an item, only for Trish to take a job as a teller in the Christchurch branch of the BNZ when she finished sixth form. I was a bit disappointed by that. But we kept in touch, and in early 1972 she applied for and got a job in Palmerston North. Eventually, she moved to Taihape to work in the ANZ and moved in with me.

Meanwhile, Wanganui Aero Work's business was expanding, as farmers clamoured to have their properties top-dressed or sprayed from the air. The decision was made to double the size of the one-man, one-plane base over at Pahiatua with the addition of another aircraft. I was offered the job, and jumped at it. Trish was lucky enough to get a transfer to Pahiatua, and on 7 July 1973 we were married in the Anglican Church in Dannevirke, Trish's home town. Neither of us were what you would call extroverted people, so it was quite a small wedding. Besides our immediate families, we had a few close friends there, including John Harding and, of course, Jeff Winter, who was best man. Jeff is Trish's partner these days, so I suppose you could say he turned out to be the best man for the job, after all.

Trish and I had a brief holiday in Rotorua, where I had my first ride in a helicopter. A local pilot by the name of Larry Bennett, with whom I had learned to fly in Wanganui, took us up for a scenic flight. I had long been interested in choppers, and had once done a quick flight in a gyrocopter, but this was my first flight in a proper helicopter. It piqued my interest.

Helicopters weren't that common in New Zealand in those days, but they weren't unknown, either. During my time in Wanganui, there had been an FH1100 parked just outside the offices of Alexander Helicopters. Mike Alexander was one of the pioneers of rotary wing aviation in New Zealand: his company, Alexander Helicopters, owned and operated a Hiller helicopter in the North Island in the early 1960s. In 1963, when he heard the news that a National Airways Corporation DC3 was missing in appalling weather in the vicinity of the Kaimai

Ranges, Mike trailered his helicopter to Matamata and presented himself to the search and rescue coordination centre—only to be told that he wasn't wanted. But several locals were able to give him an indication of where they thought they had heard the plane impact, and he decided to have a look. It was still stormy—raining and windy—when Mike spotted the crumpled wreck of the plane in thick bush on a sheer hillside. After reporting his find, Mike made over 40 flights into the crash scene, ferrying searchers and body recovery workers. I knew the story, and often had a walk around the FH1100, trying to imagine what these things must be like to fly.

As it happened, a helicopter company was starting up in my neck of the woods shortly after Trish and I were married. Hawke's Bay Helicopters was the brainchild of David Ramsay, a taxi driver who must have spent most of the downtime on his shift dreaming, as he had lots of good ideas. Dave became convinced that helicopters were the way to exploit the opportunity he had seen in the aerial application of prills (little round granules of insecticide) for the control of porina grubs. The eastern portion of the lower North Island was at that point in the grip of a major porina infestation. Porina is a moth that lays its eggs in grassland and, come early spring, these hatch into big, unsightly grey caterpillars that live underground and emerge after dark to feed on pasture grasses. The first thing you know about an infestation is the appearance of swathes of bare ground in your paddocks. Only the swift application of insecticide will save your grazing land from being turned into a barren dustbowl.

Helicopters could fly at lower levels than fixed-wing aircraft, so the insecticide prills could be applied more accurately and at lower rates. They were also less susceptible to the wind—a real selling point in the windy Wairarapa, lower Hawke's Bay and Wellington areas—so that despite the significantly higher operating costs, they were actually more cost-effective in the long run.

Dave formed a company to buy a machine and apply for a licence. His idea wasn't without support: once he formed Hawke's Bay Helicopters, most of his shareholders were farmers, along with a few local Hawke's Bay businessmen. He bought a Hughes 269B and set about applying for an air services licence. In those days, when every aspect of the New Zealand economy was centrally planned from Wellington, you not only needed a licence to fly an aircraft, but also another licence to operate an aviation business. You had to apply to a board which usually comprised three distinguished gentlemen, one of them usually from a legal background, the others often former air force pilots. They seemed to arrange things so that one of them was usually awake after lunch. Naturally, other operators would oppose a newcomer's application, and each operator had their own lawyer. The wider the geographical area you were seeking a licence for, the more operators were affected and therefore the more men in suits crammed into the hearing room. The legal profession were big fans of the scheme, but we in the industry were less enthusiastic. In this case, because we were seeking a licence for quite a big swathe of country, a long, drawn-out legal battle ensued, with no fewer than 56 days' worth of lawyers' fees payable.

Dave Ramsay had teamed up with a local engineering company, Metalform, to design an under-slung spreader bucket for applying porina prills to be carried on a hook beneath the Hughes. Bryce Easton of Metalform was an engineer with a strong interest in aviation. He owned a Cessna 180 and had lots of experience building Easton hopper boxes for top-dressing aircraft, and he'd also devised spinners for fertiliser-spreading trucks. His porina machine consisted of a hopper feeding to a motorised spinner that ensured the even dispersal of the chemical.

In October 1973, the licensing authority's ruling was in the company's favour, but it was a qualified victory: Hawke's Bay Helicopters was only allowed to perform a few of the range of agricultural tasks it had applied to do. But it was good to go, and there was plenty of demand for porina prill application.

Dave recruited a pilot: John Oscar Pyle III was an American with a couple of thousand hours' experience under his belt, including flying Huey Cobra gunships for the First US Cavalry in Vietnam. After his military service, he had joined the US Ambulance Service, expecting to fly for them, too, but had ended up at the wheel of a road vehicle instead. Feeling he was wasted in this job, he decided to move on. After a bit of research, he and his family had chosen New Zealand as a place to settle, and they had immigrated in early 1972. After a stint working as an instructor in Wellington, John had heard of the job going at Hawke's Bay Helicopters. Dave obviously did a good job of selling his vision, because not only did John Pyle join Hawke's Bay Helicopters as a pilot, he demonstrated his faith by buying shares as well.

I joined the company shortly after it had received its licence to operate. My job was loader-driver again—a step backwards, but sometimes you have to go back to move forward. I knew that helicopter work promised to provide a wider variety than fixed-wing top-dressing, where you are applying the same fertiliser from an airstrip, day after day. Still, dusting for porina was hard yakker. I had to lift 40-kilogram sacks into the under-slung spreader bucket, open them up and dump their contents into the hopper, two or three sacks to the load. I'd then start the spinner motor, give John the thumbs-up and up he'd go, lifting the hopper and flying away to the target area. There were a few minutes of rest, before the swelling flutter of the rotors would announce John's approach for a fresh load. For some reason, the spinner motor, which worked like a dream when it was new, seemed to become less reliable as it got older. Sometimes I would spend five minutes tugging away fruitlessly at the recoil starter cord before it would splutter into life, with John checking his watch and champing at the bit to get back into the air. It was a little Wankel rotary engine, and it never occurred to us that if the filter wasn't kept scrupulously clean, dust would make its way into the combustion chamber where it would wear the side-plates, causing compression loss.

Needless to say, the whole spreading operation would take place with clouds of insecticide dust swirling around the little Wankel engine beneath the hopper. I often finished a day coated inside and out with all manner of natural and manmade toxic chemicals and thought nothing of it. They were happier times: ignorance, after all, is bliss.

***

Once the seasonal porina crisis had passed, we got plenty of spraying work in the wider region. No matter how widely we ranged, though, we were usually back in Dannevirke by Friday evening. While we worked long hours during the week, John Oscar Pyle's wife Erica used to insist that he was home for the weekends, which suited me. It meant I got to spend my weekends with Trish.

One porina job in the winter of 1974 took us as far afield as Whiteman's Valley, just north of Wellington. Our visit coincided with a spell of absolutely atrocious weather—rain, day after day—making the job impossible as porina prills had to be applied in dry weather. Or, at least, it made the job impossible during the week. Every Friday afternoon, just as we were packing up to return to Dannevirke, the rain would stop and the clouds would roll away and the weekend would be gloriously fine, offering perfect conditions for flying. It made no difference to John Oscar Pyle III, however: given the choice between staying on and trying to make up for lost time on the one hand, or doing Erica's bidding on the other, there was no contest. Back to Dannevirke it was.

This happened two, three, then four weeks in a row. There were two consequences of all that lost air time. One is that after we had sat around waiting all morning every morning in an expensive hotel for the rain to clear, until it was plain there would be no window to start work, John would put the dual controls in the machine and I would get an afternoon's instruction. The other was that, predictably, the company came under pressure from the farmer, who got sick of waking up on Saturday morning to yet another weekend of perfect working conditions in which nothing got done. The company directors leaned on John, who quickly got sick of the pressure.

I got a call one Monday morning in late July.

'John, it's Graham here,' the voice on the other end said. Graham Theodore Twist was the chairman of Hawke's Bay Helicopters. 'How far off do you reckon you are from getting your helicopter ticket?'

'Oh, I'm a few hours away,' I replied. 'Why?'

'Better find someone to teach you quick,' Graham replied. 'John Pyle has just resigned.'

I ended up phoning Noel Bennett in Rotorua, and Noel very kindly agreed to come to Dannevirke. We spent the mornings spraying and in the afternoon he would put the duals in and teach me. In this way I built up my hours, and when, on 15 August 1974, I had ticked up a shade over 50 hours, we got Tom Masters from the Civil Aviation Division to come over to Dannevirke to check me out.

We did a flight test and then had a bit of a yarn about the whole business of flying helicopters. Tom was an experienced helicopter pilot and was of great assistance, giving me lots of advice on staying safe.

'Yeah, I reckon you'll do,' Tom nodded, and just like that, I had my rotary wing rating (as it was in those days), entitling me to fly helicopters without anyone holding my hand.

Like all pilot licences and ratings, it was really just a licence to learn, though. I still felt a long way away from being competent to perform as a fully fledged agricultural helicopter pilot. At this point, disaster intervened. Noel had gone back to Rotorua to have a bit of a break and a catch-up with his family. On 25 August 1974, he, his brother and his mate, the celebrated flying boat pilot Fred Ladd, went off on a demo flight in an Enstrom helicopter. Soon after they took off from a golf course where they had landed, the engine backfired, which seems to have flung the intake hose free. A few minutes later, as they were cruising above the lake at 400 feet, the open end suctioned onto the filter housing, cutting off the engine's air supply. The motor stopped. Noel performed a copybook auto-rotation manoeuvre into the water and all three men got clear of the machine before it sank. But it was winter, and the water temperature was in the single digits. Fred was the best dressed, and he managed to tread water until rescuers arrived—just one in a long string of narrow escapes for him. The others succumbed to the cold.

***

While I could fly a helicopter, I had no real training in operating a helicopter with a spray boom. In those days there was no special spray rating as there is now. I phoned Lyn Wilson, who ran a helicopter spraying operation out of Hastings. Lyn was happy for me to join him to finish my spray instruction. For a couple of weeks, we would go out onto Hawke's Bay farms spraying variegated and nodding thistles. He would divide each block in half and he would do one half and I would do the other, with Lyn's calm voice over the radio talking me through the fine distinctions between what I was doing and what was advisable. I already knew how to fly parallel spray lines using visual reference points, as you were obliged to in the days before GPS. It was a matter of learning how to handle the machine, and what in the way of hazards to keep an eye out for.

Then, at the end of August 1974, as we were flying home from a day's spraying, Lyn said: 'I reckon you know enough to survive. Go home and take care doing it. If you have any problems, just give us a ring.'

I gulped. I felt I knew just enough to be dangerous. But Hawke's Bay Helicopters had a big list of client farmers keen for a helicopter to come and do their job, and I didn't feel as though I could impose on Lyn any longer. So I got busy spraying.

It wasn't long before my lack of experience bit me in the bum. I was doing a straightforward job spraying a paddock near Eketahuna: easy terrain, not too much in the way of trees or power lines to worry about. I performed a long, straight pass spraying MCPA—a herbicide used for thistles and other broadleaf pest plants—and at the end, pulled up and wheeled around into a turn to go back and do another run. The machine didn't round out properly: it was a bit like that feeling when you're walking down the stairs in the dark and you miss one. The ground rushed up and there was an almighty bang. A drainage ditch loomed in front of me, but the machine had bounced. It struck hard again on the other side and came to rest, mercifully upright. I shut everything down and climbed out shakily. The nose cowl had dropped off and one of the spray booms was a bit bent, but there didn't seem to be any other damage. I retrieved and refitted the cowl, started up and flew back to where Grant Ward, my loader-driver, was waiting for me. He'd seen the whole thing.

'Careful with the machinery, John,' he joked, but he was white as a sheet.

We set about trying to work out what had gone wrong. We soon noticed that the leading edge of the rotors was filthy and sticky: the chemical we were spraying was sticking to the leading edge of the main rotor blades and had gummed them up, spoiling the aerodynamics just enough to compromise the machine's performance. It was a lucky escape. We both decided we'd had enough of a fright and called it a day.

***

Running choppers isn't cheap. Although there was as much work as the company could handle, Hawke's Bay Helicopters, with a working capital of $32,000, had debts to the tune of $35,000 when I took over. Some of the founding shareholders were keen to exit. Trish, who shared my faith in the company's prospects, was all for buying up as many shares as we could afford. I was a bit more conservative, so in the end, we compromised and bought out a local businessman with a reasonably substantial holding, so that we had around 25 per cent and a significant stake in the business.

As I gained experience and confidence, I started to really put in the hours and my efforts soon began to be reflected in the balance sheet. Within a year, we had our debts under control, and we were in a position to upgrade our machine, partly because we were able to get really good money for the old 269B, too. To protect the position of the New Zealand dollar against the currencies of our trading partners, everything that wasn't manufactured locally was subject to an import licensing scheme. You had to apply for a licence even to import a private car in those days—and permission was rarely given. This turned New Zealanders into a nation of backyard mechanics, keeping old cars going far longer than they strictly speaking should have, and it also saw second-hand cars fetching ridiculous prices in the famished marketplace. The economics were broadly similar where aircraft were concerned. You could get a licence to import a helicopter if you were using it for agricultural purposes, but fat chance for anything else. That meant anyone who wanted to use a helicopter for deer-shooting, for example, was willing to pay top dollar for a second-hand machine. This considerably eased the pain for us of upgrading to the far more capable Hughes 269C.

There was amazing money to be had in deer recovery in those days. One crowd that had invested in it was the Christchurch-based Edmonds family, synonymous with their cookery company, and they came knocking at the door of Hawke's Bay Helicopters, asking whether we would be interested in selling out to them. The then-HBH chairman, Barry Simmons, thanked Edmonds for their kind offer but said the shares were not for sale, as HBH was sure to rise on its own without Edmonds. We all enjoyed that comment! And the best part of it was that it was true: the company had turned right around from its debt-ridden beginnings.

There was still as much work as we could handle, too. Most of it was spraying. We sprayed gallons of chemicals—herbicide on thistles, dock, gorse and blackberry, pesticides on pasture to control porina and facial eczema, and we applied fertiliser. But every now and then, we were called upon to do different stuff.

The police approached us one day to perform an aerial search for a mob of cattle that had supposedly been rustled in the high country inland from the Bay. We spent a couple of hours sweeping the steep hills and bushy gullies. Cattle sprang from all sorts of surprising hidey-holes at the clattering approach of the helicopter, and we noted plenty of fences in poor condition. We couldn't account for every beast, but the police were able to set their minds at rest over the suspicion that a big-time cattle rustler was on the loose in Hawke's Bay.

With the tuberculosis incidence rate in cattle increasing, we participated in a helicopter-borne deer-shooting trial in the Tiraumea area east of Eketahuna to establish how far TB had spread amongst the deer population. My farmer friend John Atkins was on the local Pest Destruction Board, and he came out shooting with me. He was the shooter, of course, and worked on the principle that as long as you had lead in the air you had a chance of killing a deer. The semi-automatic 30-millimetre carbine rifle with its 30-shot magazine rattled away beside me and deer died beneath us in a hail of gunfire. What that trial showed was extensive TB amongst the feral deer population.

In a burst of inspiration, the local power board hired us to scan their lines in the rugged hills east of Pahiatua for a lines fault. It took just 20 minutes to spot the sagging wire, where it would likely have taken a day or more for a linesman on foot or horseback to do it.

And when the settling ponds for the local sewerage scheme got a bit whiffy, the contractor, Brian McConkey, called us in to take him out to a little round platform in the middle of the pond so that he could open up a valve. It was a delicate operation: I had to hover close enough to the platform so that he could climb from the machine onto the skid and then step onto the platform.

'If anything goes wrong,' I told him, 'we're both literally in the shit.'

***

These were all valuable demonstrations of the versatility of helicopters. Then one afternoon in October 1974, we got a call-out of a different kind altogether. Someone phoned Trish to say that a farmhand working out the back on a farm east of Pahiatua had injured himself, and everyone was sort of hoping we could use a helicopter to get him to the road-end. Trish phoned the wife of the farmer whose property we were spraying, and she ran the message out to us.

The farmer was happy to let us go, so off we went. It was around 20 minutes' flying, and then, in a small clearing in the dense scrub, we saw three or four men in Swanndri bushshirts, all waving at us. I got the chopper down in front of them and we were introduced to the patient. He was a short, stringy sort of bloke, and he was pale and pretty quiet. His left upper leg was bent at a strange angle and roughly splinted with a branch.

I explained to him that there wasn't room for him inside the chopper, so he would have to ride on a makeshift platform we'd rigged up on the carrying cage that was bolted outside on the skid. The poor bugger had already endured an hour's wait in pain from when the injury had happened. Anything would be better than a day's further carry. He nodded.

We made him secure and as comfortable as we could, and then we set off. It was a five-minute flight to the station house, much of it with the soaring ridges and swooping valleys of the ranges rolling beneath us. Ambulance work by air was still a novelty in those days. A small welcoming committee of farmers and wives were at the station home to meet us. If the patient had been pale before, he was positively ashen when we landed. You could tell that it would take a fair bit to get him near a helicopter again. The offer of a further flight to Masterton Hospital was turned down, with his preference to go by road ambulance from the homestead.

But, meanwhile, it showed me the possibilities. You could make a real difference in a lot of emergency situations by using a helicopter, especially if it was properly set up. Our setup with the Hughes helicopter plainly wasn't that satisfactory. Just ask our first customer!

Hovering about the new sewerage ponds in Dannevirke while Brian McConkey attempts to open a valve.

Pat Wynne attends to a patient after unloading at a hospital.

CHAPTER THREE

TO THE RESCUE

'It's getting worse,' I said to Vaughan Brown, my loader-driver, as rain rattled on the canopy and the ragged underside of the cloud skimmed just overhead. 'I reckon we call it a day.'

It was 29 August 1976. We were delivering fence posts to a high ridge on a property south of Pahiatua but conditions were rotten and deteriorating. Vaughan didn't argue. We headed for home.

After we had landed and put the chopper to bed, I went home. Trish told me that there was some concern for a plane from the Hastings Aero Club that was overdue on a flight from Nelson back to the Bay via Palmerston North. There were two people aboard. On the television news, I learned that the plane was a four-seater Piper Cherokee flown by an experienced pilot, 40-year-old Max Blincoe, who was a club member. The sole passenger was another Hastings local, 32-year-old Estelle Lyver, the mum of three young children. Blincoe had kids, too. The plane was last heard from near Ashhurst, shortly after it left Palmerston North Airport for the 40-minute flight over the Tararua

Ranges to Hastings. Having seen what the conditions were like in the area, I knew that their prospects weren't bright.

I phoned the Palmerston North Police who were coordinating the search and rescue effort and offered to use the helicopter to have a look.

'Nah, we've got it all under control,' they replied.

The following day dawned fine, and we were already in the air when the first rays of the sun touched the tops of the ranges. I thought we'd have a bit of a look for the plane anyway: we were working in the area, and it was hardly out of our way.

It's pretty standard practice in poor visibility for pilots to seek out the Manawatu Gorge to guide them through the rampart of the ranges from the Manawatu to Hawke's Bay. We flew through the gorge keeping an eye out on both sides of the aircraft. Nothing. Once through over Ashhurst, I steered south-west, along the line of the Tararuas. At about 7.15, both Vaughan and I saw a fleck of white on the hillside close to the Pahiatua Track. As we got closer, it was obvious that we'd found the plane. It was messed up, as you'd expect when a plane flies into a hillside, but still in more or less one badly bent piece.

Closer still, we could see what looked like a person lying on the ground next to the wreckage, and as I lowered the helicopter so that Vaughan could jump down from the skids, we could see that it was a human being all right, a woman, with the upper part of her body covered with a jacket.

I hovered the chopper a few metres down the hillside.

'Better see if she's okay,' I told Vaughan.

He shot me a frightened look. He jumped on and off the skid of the helicopter dozens of times a day without any qualms at all, so I knew it wasn't that. He was afraid of what he might find.

We landed on a flat area and both walked to where the woman was lying, the downwash from the rotors flapping the jacket that covered her. I took a look.

'I think she's dead!'

'Oh, Christ,' Vaughan said. For both of us, this was our first dead body.

I radioed the news of our find to Trish, and then we had a quick scout in the vicinity for any sign of the pilot. There were no houses, only a road a matter of 50 metres uphill from the crash site. It looked possible he would have made it up there if the fog that covered the hill had cleared—if he wasn't lying dead in the wreckage. It was clear that Estelle Lyver's body had been deliberately covered, so it looked like Blincoe might have survived and walked away.

When the police arrived I ventured the opinion that in low cloud, someone standing at the crash site wouldn't have known he was just downhill from the road. In the absence of any reason to head uphill, your natural instincts would send you down the hill for help.

They didn't seem too interested in my opinion.

We couldn't see any sign of him, and as there was nothing else we could do, we headed back to the helicopter and started searching downhill. We'd only really got started when the police waved me back to the wreck site and asked us to stop searching, as they had a police dog coming to see if it could track the pilot. They were afraid the helicopter would upset the pilot's scent. Vaughan and I stood around like spare pricks for a while, but it was plain that no one wanted us there. We went back to work laying out the fence. But images of that pitiful bundle on the hillside kept coming back to me, and I'm sure Vaughan was the same.

It wasn't until late in the day as we were heading back home that we heard Max Blincoe had been found. He had set off downhill as I had thought he might have, and had quickly found himself in difficult country. In trying to extricate himself from a steep little gully, he had started climbing a waterfall. He'd apparently slipped, fallen back into the stony creekbed and hit his head on the rocks. He'd likely been knocked out and had drowned in the chilly water. A searcher found his shoe downstream, and on hunting up the creek had stumbled across the body.

Over the next couple of days, further details of the tragedy emerged. Blincoe had flown to Nelson for business reasons, and Lyver hitched a ride with him to visit relatives. They were accompanied on the flight down by Blincoe's solicitor, Ron Atkinson, who was himself a former RNZAF pilot. On the day of the fatal return flight, Atkinson had looked at the weather and decided he'd prefer to take a commercial flight back to Hastings than travel in the Cherokee, which wasn't rated for instrument-only flying. Blincoe chose to fly.

Just before he landed at Palmerston North, he radioed to say that he was experiencing very poor visibility. But when he landed at Palmerston North, he seemed not only intent on pressing on to Hastings, but also in a bit of a hurry. He and his passenger were married, but not to each other: hence, it seems, their determination to get to their respective homes that evening. It's this stuff, the little ways in which emotion clouds judgement, of which aviation accidents are made.

It was raining heavily and the cloud ceiling was low. After taking off from Palmerston North, Blincoe was unable to pick up the Manawatu Gorge Road, so must have turned south to seek out the Pahiatua Track and follow that over the hills. This was a plan B commonly employed by pilots flying out of Palmerston North. Trouble was, Blincoe was too far east and didn't get as far as the track before a hillside rushed out of the gloom. He tried to outclimb the terrain but the belly of the plane struck. It bounced, the nose of the plane, including the cockpit, crumpling. Somewhere in all of that Estelle Lyver was fatally injured. Blincoe—dazed, and probably injured himself—pulled her from the wreckage. He retrieved the plane's locator beacon and tried to set it off. When that failed to work, and whether because she was still alive at that point or because he believed her to be, he covered Lyver with his jacket and set off to fetch help.

It wasn't really these details that upset me, although the image of the body on the hillside kept coming back to haunt me. The aspect of the whole terrible business that really bothered me was the doctor's pronouncement that Max Blincoe had died at around 11 in the morning—anything up to an hour *after* I had been standing with the search coordinators being told that my assistance wasn't required. It was enough to make you weep—or look around for someone to punch.

***

Six months or so after the Blincoe crash, I had my own brush with terrain. We were dropping poisoned carrots in a possum control operation, and we were having heaps of trouble with the bloody things. The product was awful, most of the carrots being rotten, so they weren't flowing well. I was trying to lay the bucket over so we could fix the problem, but that wasn't going well, either. I went to pull up again, but one of the strops became hooked over a skid. As soon as I altered pitch to climb, the helicopter canted over. It all happened very fast: the machine lay down and went quiet. The engine had stalled, the rotors bent and buckled on the ground, and the windscreen had shattered. With the Hughes out of action—we were insured and the company was sympathetic—it seemed timely to look at the possibilities of upgrading. We approached Ian Wakeling of James Aviation, the New Zealand agent for Hiller Helicopters. Ian told us he could source a pretty well preserved UH-12E for sale. This was a much more powerful machine than the Hughes: it could lift over half a ton—nearly twice as much—and, better yet, it came fitted with a factory spray rig, which was powered off the helicopter's own hydraulic system. That meant there was no need to lug the weight of an auxiliary engine around, and that capacity could be used for carrying more chemical.

We placed the order just before Christmas 1977 and within a few weeks our new machine had been landed and assembled by James Aviation in Hamilton. Ian took me out from Hamilton and showed me how to do spray turns in it, and then she was all ours.

With the extra lifting capacity and spray gear, we could not only perform our routine spraying work more efficiently, but we could also tackle a wider range of work than before. We started to get lifting work for the power board when they wanted to haul poles into the hinterland, and jobs such as lifting the roofing for the *Evening News* building in Dannevirke. As I had hoped when I took the plunge into helicopters, it all made a welcome change from buzzing backwards and forwards dropping chemicals on paddocks.

Impressed with both the Hiller and the volume of work that was available for it, we decided we could afford to expand and acquire another machine and pilot. James Aviation found another UH-12E for us, and the company hired Keith Turner, an old mate of mine from my Ag training days, to fly it. Keith and the second machine were based down the road at Masterton.

One Sunday morning soon after we'd taken Keith on, we got a call to assist with a big fire in the hills north of Masterton. The prevailing opinion was that we'd need both choppers, so I set Trish trying to track Keith down while I got ready to fly south myself. I happened to be with Hugh Barclay, a solicitor who had done all my legal work to that point, when I got the call.

'What's up?' Hugh asked.

‘Big fire down in the Wairarapa,’ I said. ‘They want us to help.’

‘Can I tag along?’ Hugh asked, his eyes lighting up.

I couldn’t see any reason why not, so we saddled up and flew down to Puketawa Station, where the fire-fighting effort was based. Meanwhile, Trish was drawing a blank trying to track Keith down. Eventually, she found someone who knew him.

‘Oh, it’s Sunday. Keith’s a God-botherer. He’ll be in church,’ Keith’s mate said.

Trish persuaded him to head to the church where he ran up and down the aisle scanning the faces of the devout for Keith.

With Keith’s input and whatever divine assistance he brought with him, we managed to get the fire under control by late afternoon. Hugh and I returned to Puketawa to refuel, and with John and Sara Atkins, the station owners, waving goodbye, we set off for home in the evening.

Ten minutes later, the station homestead was still just behind us as we battled a fearsome nor’wester. John Atkins could be seen gazing up at us from the ground.

‘We could keep going all night and we might get there,’ I told Hugh. ‘But it’s not looking likely at the moment.’

We gave up and tied the chopper down outside the Atkinses’ house, and prepared to spend the night with the very hospitable John and Sara. Hugh was pretty anxious by now, as he was due in court the following morning and wasn't confident that the magistrate set down to hear the matter would be sympathetic if he asked for a last-minute adjournment.

'You'll be right,' Sara said, and got to work on the telephone. The following morning, with the wind still roaring and fluting around the house, bouncing the chopper on its moorings, Hugh set off to use the complicated network of school buses and rides in private cars that Sara had arranged for him. He wasn't even late.

***

In late 1976, I got a call from Bill Olsen, who was the owner of an aerial top-dressing company called Fieldair on the other side of the hill. Bill was interested in buying into Hawke's Bay Helicopters, and the price he named made me sit up straight. Trish and I agreed we'd be mugs not to sell our shareholding. Funnily enough, as word got around about what Olsen was offering, there were farmer shareholders all around the district reaching the same conclusion. The deal was done pretty quickly, and, by October, Fieldair had bought everyone out.

Keith and I were kept on as the company's helicopter pilots. It seemed like business as usual at first, but soon enough, that age-old problem of transitioning from company owner to employee reared its head. When we were doing the bookings for work over the summer holiday period, Fieldair announced that they were shutting everything—including us—down for a month between 20 December and 20 January.

'But that's peak season,' I protested. 'That's when we get most of our spraying done.'

'Well, you can keep working if you want to,' they said. 'But we're off over that period, so there won't be any maintenance services. Best get any maintenance you want done before the twentieth.'

It was a very different approach to the business to the one we had taken, and which had worked for us up to the buyout. We looked at what farmers wanted first and foremost and then we arranged our operations around that. Fieldair seemed to come at it from the diametrically opposite direction. Meanwhile, just about everyone I knew was heading offshore to all kinds of exotic locations. Helicopters and pilots were in big demand, and there was big money to be made. Quite apart from the fact that I was about to have a family, I had no real desire to go overseas to make my fortune. In fact, the way it turned out, the exodus of young chopper pilots worked in my favour, because there weren't many of us left kicking around at home.

One evening, a bloke knocked on the door who introduced himself as Robert Kay. He was doing a bit of touring around getting to know who was who and who did what in helicopters in the North Island, because his company—Nelson-based Helicopters New Zealand—was looking to hire someone to fly for them in the Taupo area, where they had just landed a couple of big contracts spraying forestry blocks and assisting with the construction of a new skifield on Mount Ruapehu. Their regular pilot, Rusty Nairn, had recently crashed and sustained burns that were going to keep him out of action for some time.

'What machine do you use?' I asked.

'Well, we've just got a new Lama, and we've ordered another. You know it?'

I knew it, all right. The Aerospatiale 315B Lama made even the Hiller look like a toy. It could lift an imperial ton. Flying one of those things would be a dream come true.

Again, Trish and I had little difficulty deciding it was the right move. So, in March 1977, I tendered my resignation to Fieldair and flew to Nelson to begin my induction. It was a time of change all round, really. Trish and I had our first child, a baby boy, on 7 September 1977. Mark was followed a bit over a year later, on 31 July 1978, by Hamish.

***

Helicopters NZ Ltd (HNZ) was founded in the early 1950s by a Timaru grocer named

Doug Shears, who had been a pilot during the war and who could see in the emerging technology of rotary wing aircraft an enormous potential to be useful in New Zealand agriculture. He had several goes at getting a licence, but it wasn't until 1955 when he teamed up with a small-time accountant named Alan Hubbard that he managed it, and also bought a Bell 47. The rest, as they say, is history: the company's fleet expanded, and they became the dominant provider of helicopter services to all manner of industries, not only farming but also forestry, mining, oil and gas exploration, and tourism. Since 1963, it had been based in Nelson, where it was managed by Robert Kay and Peter Tait. It was a dynamic company, with an attitude that no job was too hard, and they put a heap of effort into me.

It was Peter who chaperoned me around my first few jobs. We did some lifting: one of them was a supply job to the lighthouse on Stephens Island, which involved carrying a considerable load onto the island from a ship. The Lama chewed through a heap of fuel—nearly five litres a minute at low altitude—but by carrying spare fuel aboard, we were able to triple our operating time from one hour to three. Peter taught me heaps about rigging up loads. We did a big spraying job on Molesworth Station in Marlborough, and assorted others even further afield, which saw me do a meandering tour of the South Island meeting some of the real characters of the industry. Russell Gutschlag was a bit of a legend in the lower South Island. He had learned to fly in Tiger Moths, ran his own top-dressing operation, and became only the fourth person in New Zealand to receive a rotary wing rating when Sir Tim 'Hurricane' Wallis paid for his training. He then became one of Wallis's most successful venison recovery pilots. One day when I was working with Russell, we got called to lift a Piper Cub that had made an untidy landing in a riverbed when engaged in deer recovery. Russell taught me how to configure the little plane so that it didn't try to fly as you were carrying it under the helicopter. Others such as Neville Cox, another pioneer who was HNZ's pilot in Geraldine, taught me heaps, too.

On the rare weekends we were able to take some time off, Trish and I were in the habit of heading up to Taupo to do some waterskiing or fishing with friends. Taupo was a bit of a backwater in those days, and so we were able to buy a little one-bedroom bach for $12,000, where we based ourselves on these expeditions. That was to be our home when we moved there to start operations for HNZ.

The Lama wasn't quite ready by the time the construction of the Turoa skifield kicked off in early 1979, so when I arrived in Taupo ahead of Trish and the boys to start our new life, I was flying HNZ's Hughes 500. My job was mostly ferrying construction workers on and off the mountain, and my point of liaison with AHI Harvey, the construction company, was a bloke named Dave Johns. Dave was known as a highly competent operator, but a hard man. Plenty of people didn't see eye to eye with him, but as is so often the case, it all boils down to work ethic. I think Dave decided pretty early on that I was a can-do sort of guy rather than the kind who looks for excuses not to fly, and he appreciated that.

Not that there weren't plenty of excuses available. Mountain weather is quite often far from conducive to the operation of helicopters: low cloud, rain, hail, sleet, even snow are common, not to mention high winds that interact with the terrain to set up all manner of interesting turbulence patterns. I was often flying in atrocious conditions, especially once the Lama arrived and the emphasis shifted to lifting. It was pretty difficult to say 'nah, the weather's too dodgy' to the driver of a concrete mixer who had just arrived at the pad with a full load of mud to be ferried up the hill. Some of the most exciting times I can remember were spent delicately lowering a hopper of concrete to a bunch of dim figures clustered on the ground as cloud intermittently blew through to obscure everything. Sometimes it was like hovering in a glass of milk.

There were some lighter moments, too. When it was snowing lightly or when the wind was very high, the Lama used to build up quite a charge of static electricity. One day in particular, I was delivering a hopper of concrete to a bunch of guys on a scaffold. As I lowered it gingerly toward them, one of them stretched out a hand to steady it and a brilliant blue spark, fully a foot long, leapt from the nearest point of the hopper to his hand, flinging him backwards. Someone else stepped up to try again. Another spark: same result. I pulled up and began another approach, but this time I could see everyone on the scaffold scrambling to get away from the approaching hopper. From my vantage, it was a hilarious spectacle. Hovering a helicopter when you've got a serious case of the giggles isn't something they teach you in flying school! Eventually, I worked out that if I gently nudged the hopper against the scaffold before it was handled, the charge was earthed and no one got more than a mild tingle as they grabbed it.

By now, Trish and the boys had come through from Dannevirke, and we were installed in our little bach. It wasn't that great at first. Mark was a difficult sleeper—he'd usually need a solid hour or two touring the block in a stroller before he would settle—and given the hours I was working, the situation wasn't ideal. I ended up setting up a bed in the garage so that I could get an uninterrupted night's sleep and slip out in the morning without waking anyone.

***

The Lama's unrivalled capabilities meant it was frequently in demand for jobs at all manner of exotic locations. One saw me travel to Wellington to lift some massive air-conditioning units to the top of the new Challenge Tower on The Terrace, and another to lift six air-con units to the roof of a new building at the Tomoana freezing works in Hastings. We were always being called upon to lift power poles, and in early 1980 I did a big job shifting nearly 1000 sewerage pipes from railways wagons and laying them alongside a trench between Rotorua and Ngongotaha.

Another feature of this aircraft is that you could take a front seat out and create enough room for a stretcher to be laid down inside the cockpit. I hadn't been working at Taupo very long before we got the opportunity to test this capability. One afternoon around lunchtime, we got a call to go and pick up a man whose leg had punched through a crust of earth over a fumarole close to the Ketetahi hot springs on the northern slopes of Mount Tongariro. He was very badly burned, and was lucky a park ranger had happened to be walking the track just behind him. He found the man and his distraught tramping companion and raised the alarm. The weather was awful, and getting worse, but I thought I would have a go at getting in there, anyway. The same weather conditions meant it would have taken a good seven or eight hours for a ground party to walk in and walk out with the victim on a stretcher.

I flew to National Park headquarters and collected a local cop by the name of Rod Kitto and Doug Wilson, a park ranger, and we headed for Ketetahi. The cloud ceiling was right down, and it was drizzling steadily. My passengers were pretty apprehensive as they stared out at the dim, grey earth passing below.

Things got worse around the northern side of the mountain. The wind had picked up, and you could hardly see a thing, but we managed to follow the track to where it breaks from the bushline. The springs are a little higher up the mountainside, and we soon located a small knot of people around the victim, who was under cover on the rocky ground. I identified a suitable spot to set the chopper down and did so gently, then we loaded up the victim, a 38-year-old Aucklander. His left leg was red and raw from mid-shin to foot, and he was in a lot of pain. Within 20 minutes of picking him up, we delivered him to an ambulance at National Park and he was on his way to Taumarunui Base Hospital.

'Shit, that was hairy,' Rod Kitto grinned, once we were safely back on the ground.

'You looked pretty calm,' I told him.

'That's because *you* looked so bloody calm!' he replied. 'I was watching you the whole time, and if I'd seen the slightest trace of anxiety on your face, I swear I would have shat myself!'

Completing a lifting job in the Lama at Fletcher’s Mill, Taupo.

Attending a motor vehicle accident.

CHAPTER FOUR

LIVES ON THE LINE

The capability of the Lama soon saw us in heavy demand all over the North Island. I had a contract up in the Far North, a big day spent lifting 230 kauri logs out of the Russell State Forest for the Forest Service and, in February 1980, I also got a spraying job just outside Herekino, west of Kaitaia. Dean Sparks was with me as my loader-driver.

I was performing what ought to have been a routine spraying run at about nine in the morning when I felt the machine shudder and try to lean over to the right. There were sparks, and the sickening realisation I had tangled with a power line.

I had hooked up a wire once before, when I failed to notice an electric fence a farmer had strung across a gully on a property I was spraying in the Turakina area near Wanganui. He had neglected to tell me about it. I found myself with one end of the wire firmly tangled with the spray boom and the other end just as firmly anchored to the ground. I could maintain translation—forward movement—but could only go around in circles, like a kid's trainer aircraft. Luckily, I was able to radio Dean (who was there for that one, too), and we found a level patch of ground to land on to untangle the wire from the skid.

The important thing is to make no sudden control movements, and to keep the attitude nice and level. I worked out that the power line on this occasion had fouled the high frequency (HF) radio aerial mount on the tip of the skid and was snarled between the chopper's chin and the skid. No amount of manoeuvring would allow it to come free. I tried reversing, but the wire wouldn't budge. I tried moving sideways: there was a shower of sparks, but no sign that it was going to work free. So I tried moving forwards to see if I could break the thing. I changed pitch and carefully eased forward: I could feel the wire tensioning up, and then there was a sudden lurch. One of the wooden corner poles had broken. Now I was dragging a wire and a few hundred kilos of tanalised pine log, too. I eased forward again, with the same result: another pole gave way. I was wondering how much the national grid weighed, and was afraid I was about to find out.

By the time I had two poles and a 200-metre, sagging skein of wire attached to the chopper, I could no longer move forward. I didn't want to lower the collective (the collective lever controls lift by altering the pitch of the main rotor blades, and is used to command the helicoper to rise or fall) and try to set down, because that would bring the rotors dangerously close to the wires, and besides, I had no idea whether the wires were still live or not.

I raised Dean on the radio and explained my predicament. Then it was a matter of waiting for the cavalry to arrive, in the shape of the loader running across the paddock below.

Once I had descended to a hover as low to the ground as I dared, we had the far from trivial matter of how to prise the wire off the skid, given it might well still be carrying a few hundred volts of electricity. Dean eventually found a dry fence batten lying around, and used this to lever the wire off. I was free! I set down at a safe distance from the wreckage of wire and poles and climbed out, wiping the nervous sweat from my forehead.

The farmer, who had seen the whole thing, approached.

'I think we'll call it a day there,' he said. 'Come up to the house for a cuppa.'

There was, of course, no power, but we had the kettle bubbling away merrily on a gas hob when there was a loud knock on the door. The farmer opened it to find a pretty pissed-off-looking bloke in a Bay of Islands Electric Power Board uniform standing there. He was holding an item that Dean and I immediately recognised as the HF aerial that Dean had rigged up on the right-hand skid.

'That your helicopter that hit my lines?' he barked.

We all looked at one another and shrugged, the very picture of innocence.

'Don't know anything about that,' I said. 'Where's your line?'

He looked as though he might explode, and he brandished the aerial.

'I've got a piece of helicopter here,' he said, showing us the broken end. 'I'm going to walk around your helicopter, and if I can match this up, you'll have to pay to replace the line and poles.'

We maintained our poker faces, and after a moment he whirled around and marched off.

'Hey, mate!' the farmer yelled after him. 'I haven't got any power. D'you reckon you can get on to fixing it, please?'

***

That wasn't the worst brush with catastrophe I had in those days.

Dave Johns, who was the foreman in the construction work on the Turoa skifield, was also a general contractor. He was known to be a bit of a genius with a little earthmoving machine called a bobcat, and one day he was approached with a tricky proposition. There were a bunch of airstrips high in the Kaimanawa Ranges which, due to their remoteness, had been constructed with hand tools. A local aviation company operator had just secured a licence to use a few of those strips, and was keen to have them improved. He asked Dave if he could think of any way of getting his bobcat up there, and Dave thought of helicopters and of me.

Helicopters New Zealand had a long aluminium beam that it had built in order to lift big loads such as the huge chairlift bull wheel we had installed up on Turoa. The idea was that you could attach a helicopter at each end, and that would halve the load each machine had to lift. It was a very dicey operation, of course, because if one helicopter strayed into the wash from the other, then its flight performance would be compromised and disaster would likely ensue. But we had managed a few lifts with it, and it seemed to be the answer. It was stored at Taupo. We figured that if we stripped the bobcat of everything we could—bucket, arms and so on—we could get its weight down to about 1100 kilograms, about 100 kilograms over what the Lama could lift alone. If we used the beam with a Hughes 500D (belonging to fellow chopper pilot Toby Clark) on the other end, each machine would only be lifting 550 kilograms, which was within the lifting capability of the Hughes.

So we gave it a go. Things were going quite smoothly—we were aloft and high in the Kaimanawas, Toby and I talking to one another constantly and keeping everything under control, when I felt a slight lateral tug through the controls. I glanced out the window and saw the beam starting to fold in the middle, which was pulling the two helicopters together.

'Christ!' I yelled, and I pushed a button intending to open the hook which attached the bobcat to the beam. Instead, I hit the button attaching the beam to the Lama. I knew what I had done in an instant, but it was too late. The entire 1.1-tonne load was transferred to the Hughes, which wasn't rated to lift that weight, let alone at an angle. I watched as the load whirled the Hughes around like a toy and the whole assembly plunged 100 feet to earth.

I was sure I had killed Toby. I was already wondering what I was going to tell his wife.

Fortunately, the bobcat struck the hillside first, which momentarily unloaded the Hughes. Toby managed to drop it and get clear. The bobcat rolled two or three hundred feet down the slope into the creekbed at the bottom.

We returned to base and had a bit of a debrief, both of us ashen-faced. Toby would have had no time to do anything but react, even if he'd had any control at all, which he didn't.

The next day, which was cool and with a favourable wind, both of which factors improve the lifting capacity of a helicopter, I managed to heave the bobcat up to the ridge using the Lama alone. Dave got the airstrip tidied up.

But I am still haunted by that sickening moment when I realised I had hit the wrong button.

***

Not long after we arrived back in Taupo from the job in Northland in early 1980, I got an emergency call towards the end of a long day's spraying. Someone was in trouble in the river just above Huka Falls. Details were a little bit sketchy, but it seemed it was a man who was fishing from a rocky island that was fast disappearing as the gates controlling the level of Lake Taupo had been opened. I had Mark with me, and I had to dump him at the airport. He was only four, but he nodded solemnly while I told him to stay put until someone came to look after him. I wasted no time stringing a scoop net below the Lama and getting into the air. I radioed the tower and asked Roy Carmichael, who was on duty, to get someone to go and fetch Mark. Meanwhile, I was going full noise towards the falls. It was only a matter of minutes before the sparsely grassed pumice plateau beneath me was slashed by the river in its bushclad gorge. There was a group of people on the riverbank and firemen working frantically to rig a rope at the top of the falls—and then I saw the man himself, pretty much in mid-stream, his silver hair quite distinct against the green water.

It was like one of those bad dreams where you seem to be powerless to influence events, or to move quickly enough. I dropped toward the river, but the rock on which he was standing was completely submerged, and I saw him lurch with the force of the water. He slipped under and was dragged off toward the falls at a frightening speed. I chased him and dipped the net at the head of the main cataract, but it came up empty. He was gone.

I flew over the white tumult of the falls and dipped the net again in the deep, blue pool at its foot. Nothing. I tried again and again over the next half hour—a jetboat upriver arrived to help—but there was no hope by this point. I briefly saw a set of waders, but there was no sign of the body.

It was all in vain. It didn't really affect me at the time, but once I got back to the quiet of the hangar and was putting the helicopter to bed, the whole ghastly thing just kept replaying before my mind's eye. Then the if-onlys set in, too. If only I had got the call a minute or two earlier. If only I had pushed the machine even harder and flown a bit faster. If only I had broken the rules and flown lower...

Details emerged over the next few days. The dead man was Max Pearl, a 56-year-old doctor from Auckland, a keen outdoorsman. He had waded to the rock from which he was fishing through knee-deep water, and had been so preoccupied he hadn't noticed the sudden, insidious rise of the river at first. By the time he did, he was stranded. Other fishermen raised the alarm, and the police and fire service had several goes at getting a rope to him. One cast had actually landed at his feet, but he couldn't grab it.

I was still brooding on this episode a month or two later when I got an almost identical call. A woman was in the river above the falls. This time, I didn't stuff about with observing CAA regulations about minimum heights: I just hauled ass to the river, thundering over town with the scoop net dangling beneath. She was clinging to a log at the riverbank, and it was a relatively easy matter to dip the scoop net into the water next to her. She clambered in and I lifted her to safety. This time, the circumstances were very different. She had a history of mental health problems, and had been seen to jump from the Huka Falls bridge. There was a fair bit of speculation that her decision had been influenced by the publicity surrounding Dr Pearl's death. What made it worse is that some time later she tried again. After the second attempt, I asked her if she could let me know next time she planned to do it, so that I didn't have to push the helicopter so hard to get there.

***

Dean Sparks had a real gift for electronics. He was a bit of a genius, a seat-of-the-pants improviser, and when we heard about radio direction finders, he was sure he would be able to wire one into our machine. So we got hold of one and Dean did the honours.

Our first call to use it came one afternoon when we were heading back to base from a job north of Taupo. We picked up the unmistakable beep of an emergency locator beacon. We flew back to the airport, took out the front seat to make room for a stretcher and set off in search. Using Dean's rig we were fairly quickly able to find the source of it: it was the mangled remains of a Hughes 500 belonging to Joe Collins. With the rise of the market for venison, the shooting of deer from helicopters, which had been a lucrative industry in the South Island for some time, began to take off in the North Island in the early 1970s. Inevitably, given how risky that particular business is—using a helicopter to chase agile animals in bush and rugged terrain and often marginal weather—so did the number of 'incidents' involving choppers.

Joe had suffered engine failure over an area of low vegetation, but had managed to get his machine to a relatively clear patch before it rolled over several times on landing. Apart from no blades or tail boom, it was more or less in one piece. As we neared, we could see all three occupants, two of them at least ambulatory—the paramedic's fancy word for walking around—and one of them lying on the ground. We landed close by and jumped out to see what the story was.

All three men had a fair amount of blood on them, but given the business they were in, it was unclear whether this was their own or deer's blood. One of them—Joe's brother—was making a fair bit of noise, moaning and groaning and telling everyone how badly injured he was. Joe himself was quiet but seemed alert and lucid. The other was warm, but apparently unconscious. Applying the John Funnell Patent System for the Classification of Death and Serious Injury, we judged the man on the ground to be the worst casualty. The noisy one was alive, judging by the noise. Joe Collins was pretty good, I reckoned.

We didn't have anything much in the way of specialist equipment, only a homemade stretcher about 2 metres long and half a metre wide. We secured the guy on the ground on this and loaded him in the Lama, but that left only three seats for four men. As pilot, I had a reservation. Dean got his own seat, too, which meant that we had to stack the other casualties two-high in the back, with Joe perched on his brother's knee in the seat behind mine. Joe took care to poke the men's rifles in the back, too.

I radioed Trish and told her to contact Taupo Hospital and tell them that we had three casualties of a helicopter accident aboard, all injured, one of them seriously. We lifted off and I pointed the nose towards Taupo.

As we flew fast and low towards town and the hospital, a dead weight suddenly settled on my left shoulder. It was Joe's head lolling forward. That shoulder was connected to my collective hand, of course, so it was making flying a bit difficult.

I reached my left hand up to give him a shove. I placed the palm on the crown of his head and pushed, and there was a peculiar, slippery sensation as his scalp peeled back from his brow.

'Jesus!' I yelled. 'Get him off me! I can't fly with him all over me!'

As we approached the landing site (the grounds of Taupo Hospital, just outside the main Accident and Emergency entrance), we could see a huddle of white-coated nurses waiting for us, along with a fire engine that had helpfully showed up in case we crashed and burned at the hospital. I got the machine on the ground and shut it down. We climbed out as the nurses raced forward. They loaded the unconscious man on a gurney. Joe and his brother—the former back from his faint—got out stiffly, looking as though they'd just done a shift at the freezing works, covered in blood from head to toe. They automatically pulled their rifles from the chopper—like tradesmen of every description, deerstalkers don't like to be separated from their tools—and began walking towards the hospital.

'Hang on, Joe,' I called. 'You can't take those in there.'

Joe looked at the ex-military semi-automatic FN rifle in his hands and saw the truth of what I was saying. Reluctantly, the two of them gave me their guns and I poked them back into the chopper.

We accompanied the wounded as far as we were able. Then, once it was clear we were of no further use, Dean and I mounted up again and flew back to the airport to put the seat back in, clean out the blood and generally set things in order. We landed to find all hell had broken loose. All kinds of people wanted to talk to us—police, ambulance, fire and plenty more besides.

'Why weren't we told about this accident?' was the most popular question, often spoken in tones of personal injury, as though we had committed some kind of turf violation.

'Well, we were just flying along and we heard the beacon. We were only doing what we thought was right...' I tried to explain.

***

There was some fallout from the whole incident, although probably not as much as there should have been. All three men made a full recovery, but this fact seemed to have been lost on the St John's people. I tried to tell them that the helicopter was ideally suited for this kind of work, and given the speed with which you could get accident victims to hospital care, it could dramatically improve outcomes.

They weren't interested. Their operation was as good as it gets, and there was no need to involve new-fangled gizmos like helicopters.

'As I understand it,' I said to more than one St John's bloke, 'you people started out scraping bodies off the battlefield and putting them in handcarts. These days you use vans with engines. So there must have been some point in your history when your organisation was open to new technology...'

This argument didn't seem to cut it.

Laurie Burdett, who was the charge nurse and manager at Taupo Hospital, seemed to be far more impressed with us. Laurie in particular wanted to know what training we had had, and was gobsmacked when we shrugged and said: 'None.' She looked into the possibility of sending a nurse out with us whenever we had a call-out. And the fire service, who were accustomed to being on the front line in emergencies, saw the possibilities in what we were doing as well. The very community-minded local fire commander, Peter Smith—known to all as EPH, because of his initials—called me and asked me what we were doing for medical staff. When I told him I presently didn't have any medical staff to call on but thought I might have nurses along in the future, he offered to assign a fireman with first response experience to go along as well. Richard James and John Hare were a couple of the guys who I worked with early on, but there were others, too.

So that was the way it started out. When we got a call-out, I would land outside the fire station, load a fireman and we would fly to the hospital to collect a nurse and her equipment. The firemen were always booted and suited for the kind of work we were doing, but the nurses were in their hospital uniforms—the little rubber-soled sandshoes that they use to squeak up and down lino hospital corridors, cardigan and immaculately white smock. It was the smock that was plainly unsuited to this kind of work, as it was often cut quite high above the knee. The nurse would perch on the seat behind me, and each time I turned to talk to her, I often had difficulty looking her in the eye.

After two or three call-outs, Peter asked me how the whole arrangement was working. Fine, I told him, although I confided that the standard-issue nursing uniform was a poor fit with helicopter work, not least because of the danger of distracting the pilot. Peter chuckled, and offered to supply the nurses with overalls and boots. That made far more sense, even if it made for a far less pleasant view.

***

We began to get call-outs to accidents and emergencies quite regularly, as many as one a week. Inevitably, the local doctors got wind of what was going on, too, and they decided that they should be involved. After all, a fully qualified doctor could render assistance at a much higher level than the first aid administered by a nurse or fireman. So the doctors put together a roster and for a while I would lift off from the fire station or hospital with a fireman and a doctor aboard. But we noticed that each doctor would do two or three jobs and then declare themselves unavailable. When I asked why the novelty was wearing off, Peter Battersby—my GP these days—explained that while they all thought the work was important and even, at times, enjoyable, there was a downside. The doctors all had their own practices, and each time they were called away at very short notice, spending two to three hours flying to the scene, stabilising the patient, flying to hospital, administrating the handover to hospital staff and then flying back, they would come back to a waiting room full of grumpy patients and would have to work overtime to clear the backlog.

It was soon back to nurses and firemen again. I had a paramedic by the name of Spencer Wallace come along a few times, but then one day a young ambulance driver by the name of Pat Wynne arrived in town from Napier. Pat was slightly built with lots of dark hair and a shaggy beard (as was the fashion in the seventies). He was good fun but, as I soon learned, completely dedicated to and focused upon saving lives, and not too fussy about breaking rules and regulations if they were getting in the way of core business. I remember one job I did with Pat quite early on where we attended a crash on the Napier–Taupo highway. By the time we arrived, there was quite a crowd of people standing around. Pat elbowed his way through and knelt to examine a woman who was plainly very badly injured. He quickly reached the conclusion that she had a serious chest injury, so he produced a set of shears and began cutting away the woman's clothing from the hem of her jumper to her collar, clipping her bra through as he went.

'You can't do that here!' came the aghast voice of a woman in the crowd.

From where he was kneeling and without looking around, Pat simply made a flat, sweeping motion with his right arm, a gesture that spoke louder than words: 'Don't bother me with that crap. There's work to be done.' That was Pat's attitude, right there.

Pat's training and his dedication meant we were quite an effective team. He was ready to go at a moment's notice, and once we were on the scene, he was all action.

***

In the days before pagers and cell phones, getting hold of someone when they were out and about was potentially a real problem. Our method was for me to keep Trish informed of my movements, so that if a search and rescue (SAR) call came through, she would know which farmer or forestry company manager to phone to get hold of me. That was fine if I was working. However, trouble arose from time to time when I was simply in town doing the shopping, or something like that. I might go to a shop and then decide I may as well drop into the chemist. Trish would have picked up a call and dashed into town to find me. After racing up and down the streets looking for my car, she would then try to figure out where I might have gone next. Often the first inkling I would have that there was something in the wind would be when I would see her driving apparently aimlessly around town.

So it was one day when there was a terrible head-on crash at the set of corners on the Desert Road known as the Three Sisters. When Trish found me, I raced to the airport. Pat had already got the word, and he was there with his bag of tricks. We were in the air in short order, and soon on the scene—the usual gut-wrenching sight of twisted metal and broken glass sprayed across the roadway, the smell of petrol and oil and hydraulic fluid.

Pat assessed the injured and reckoned that one—severe head and internal chest trauma—was critical: the rest could wait for an ambulance. So we loaded the victim aboard and got aloft again.

Soon after taking off, Pat came on the intercom.

'This guy's bad. We need to get him to Waikato urgently.'

We thundered towards Hamilton, the Waikato River and the lakes behind the dams gleaming in the last rays of the sun. I don't remember how we got word through to Waikato Hospital. We must have radioed Trish and got her to phone ahead. It was arranged that we should set down in a park just down the road from the hospital. It was pretty much dark as we swept over Lake Rotoroa and Frankton, and we could see the landing zone ahead, a blaze of headlights and the revolving red lights of an ambulance and fire engines.

'He still with us?' I asked Pat.

'Just,' he replied.

'Only a couple of minutes,' I told him.

No matter how urgent the matter at hand, your job as pilot is to operate the machine safely. If you rush things or push too hard, you might end up making bad decisions and jeopardising the life of more than just the patient. And if it happens that you're too slow, then, well ... que sera, sera, as the song goes.

We touched down and the ambulance crew raced forward and swarmed all over our patient. He was soon loaded into the ambulance, and it sped off toward the hospital. There was nothing to do other than climb aboard and fly back to Taupo in the dark—a routine task these days, but a bit of an adventure back then. Once on the ground, I had to put the machine away, fill out the paperwork and generally get ready for the following morning. It was after midnight when I finally got to bed.

At around seven in the morning, I had a phone call.

'It's Dr Hyde here,' a young male voice said. 'Is that John Funnell?'

I admitted I was.

'I'm one of the intensive care specialists up here at Waikato, Mr Funnell,' he said. 'I was wondering whether you'd be able to come up to Hamilton to have a bit of a chat with us.'

Christ, I thought. What have we done now? It seemed we were always in trouble, those days. The only variation was how deep.

'Is there a problem?' I asked.

'Not really,' Dr Hyde said. 'We'd just like you and Mr Wynne to come up and see us.'

So I rounded up Pat and we drove up to Hamilton and parked in the grounds of Waikato Hospital. We found our way to the admin section, and after a bit of a wait a young man—a bit younger than me—appeared, smiling affably. He didn't look as though he was about to deliver a bollocking. We shook hands. Rowan Hyde showed us into his office and sat us down.

'How's the patient from last night?' I asked, fearing the worst.

'Oh, he's fine,' said Rowan. 'Or, at least, he's not fine. He's still critical, but he'll probably pull through just fine. That's sort of what we wanted to talk to you about.'

Pat and I exchanged glances. Here it comes, we thought.

'We would never see a patient with those kinds of injuries here,' Rowan said. 'Or not from out of town, anyway. By the time we got an ambulance to them, worked on them and tried to get them in, they'd be gone. No chance, really. So we can see the value in what you're doing, what you did last night.'

Pat and I looked at each other again. Perhaps we weren't in trouble, after all.

'And we have a situation with the bad cases in the T hospitals—you know, Taupo, Tokoroa, Taumarunui, Te Kuiti, Thames and Whakatane,' Rowan went on. 'Quite often we have to transfer them here, and by the time we've sent a vehicle out, stabilised them to the point where they can travel and then driven all the way back, it can be six to eight hours down the tubes. A whole shift. So a few of us were having a talk, and we were wondering whether we could use the helicopter for that sort of thing. Transfers, and so on.'

I shrugged. 'Fine,' I said.

The necessary arrangements were made, and soon we had added hospital transfers to our portfolio of tasks. I'd fly up to Waikato and land on the tennis courts outside the nurses' home (the nurses were, generally speaking, either working or sleeping and not likely to be playing tennis). I'd collect a medical team and we'd head off across country to whichever of the outlying hospitals required our services. I quickly found that if I grabbed a bag or a clipboard or some other piece of medical paraphernalia, I'd be able to tag along as the medics did their work. No one asked any questions. I watched everything with a keen interest, and soon I was a bit of a dab hand at setting up heart monitors and medical equipment like that. So I would follow the patient from the emergency room at the provincial hospital right through to the intensive care unit at Waikato, and often I was actually quite useful as the team scrambled to stabilise the patient.

There was a steady flow of transfer work, and it soon fell into a recognisable pattern. The so-called T hospitals had only a skeleton surgical crew, so were ill-equipped to deal with very serious cases. But they tended to apply what they called a 'stay and play' policy, where they would do their best to manage the hard cases themselves, purely in order to keep their skills up (and most probably to justify the presence of surgical staff at all). But come the weekend, your sole charge surgeon was likely to want some time off, and would make arrangements to have their patient delivered to Waikato. I most often received the call at around five or six on a Friday afternoon. So a normal Friday for me in those days would be: up at 4.30am to get to the airport at five. I would be on the job—mostly spraying—at 5.30 (a bit later in the winter), and would work solidly until midmorning, perhaps 10.30 or 11. I'd then take the helicopter back to base and give it a decent wash, and if there was nothing else on, I'd have the afternoon off. Quite often, though, there was time to fit in a small lifting job or the like in the afternoon, and then I'd get the call to do a hospital transfer. I would collect the transfer team from Hamilton, fly to the hospital concerned, wait or assist as the patient was prepared for the flight, and often not arrive back at Hamilton until 10.30 or 11 at night. Then, my duty done, it was back to Taupo, clean and tidy up the helicopter, do the paperwork, and then home, often around midnight, ready to get up at 4.30am to start again.

An unhappy day: auto-rotating into Happy Valley.

CHAPTER FIVE

ECHO FOXTROT ZULU

In 1984, the Ministry of Transport served notice that it intended delicensing the aviation industry. Gone would be the hated, costly ritual of applying for an air services licence, but gone, too, would be the protection from competition that a licence offered. Helicopters New Zealand considered their position, and decided they would scale back their sphere of operations and concentrate on the lucrative offshore work with the oil and gas industry. The first we knew about it was when the company operations manager did a tour of the various HNZ bases and made a detailed inventory of all of our plant and equipment. A few days after he had been through, the company called a meeting of all its North Island operators at Taupo. We each had two options, they told us. We could either buy our plant and machines, or we could look for another job.

Trish and I knew we had a really good operation in Taupo, and there was every reason to be confident that the opportunities available would continue to grow. Bob White, who was in charge of doing the deals, offered us a very fair price on the machine we were using. The

Lama had lately been replaced by a flash new Aerospatiale AS 350 Squirrel, and it had very low hours on it. White was prepared to let us buy our operation without charging any kind of premium for goodwill. Trish and I decided to accept, and we mortgaged the house to finance it. It was exciting and terrifying at the same time, because the economics of the commercial helicopter business involve lots of very large numbers. It's possible to make really good money, but the whole time you're doing it, you're forking out sums that are just as large both in operating your machine and, especially, in maintaining it. We devised a scheme where we would set US$300 for every hour we flew aside against future maintenance bills: this reserve was the difference between operations like ours, which succeeded, and others, who found themselves crippled by the bill for a rebuild against which they had made no provision. Later, Trish had the brilliant idea of using this money to invest in an asset, against which we could borrow when we needed the money for our maintenance requirements. After looking around, we bought the Suncourt Hotel on the lakefront in Taupo in April 2000. Trish had the idea that tourism would be a growth industry in Taupo; she was right, but I'm sure even she is amazed at how right she was. We bought in before property prices in Taupo went through the roof.

Soon after we went out on our own, we were fortunate to land a contract spraying the vast Kaingaroa forestry blocks for the Forest Service. We decided that the most efficient way to do it would be to spray basically around the clock, and we got special permission from the Civil Aviation Division to use a method we had devised to do it. We would do a walk-through with the forestry guys of the area we intended spraying, and they would remove any spars or projections from the treetops that might snag an unwary chopper pilot. In the evening, they would set up a ute with a tall pole with a strobe light on top of it at either end of the row where we were to begin. When I arrived at night, I would use these beacons to line up on the row and then do a run from one light to the other. I had a big array of powerful lights mounted at the front of the machine to keep visual contact with the ground. As I left a beacon behind, it would be shifted along to the next row. It was very precise—I don't think we were much less accurate than we would be today using GPS. The other advantage was that it doubled our flight time. I would spray from around 9pm to daylight, then John Fogden, a local boy whom I had employed to fly for me, would take over and fly till nine or ten in the morning.

It was going really well, until we came a cropper. We landed a job up on Whakapapa skifield, and I was trying to balance this with night spraying. I was based at the Chateau, and one morning I climbed into the machine and ran through my usual checks. I started up, and we readied the first load to fly up the mountain.

I had lifted off and was approaching the top of the Bruce Road when the engine died. I dropped the load, and used the energy left in the rotors to guide the chopper into Happy Valley. I didn't do too bad a job of the landing, but the main rotor blades struck the hill and were badly damaged.

Upon investigation, I found that there was an awful lot of air in the fuel tank. In the Squirrel, there is a small panel—about 7.5cm x 5cm—with a couple of rows of lights that are intended to warn you of any impending failure. The lights are too bright when you're flying at night, so you can dim them. The trouble is, if you neglect to un-dim them, you probably won't notice them in the daylight. I plainly hadn't noticed them.

I felt sick to the pit of my stomach. Trish and I had poured everything into that helicopter, and we had no equity in it to speak of. And without a helicopter, we had no means of servicing our huge debt. It looked like our business would fail and we would lose our house to boot.

In desperation, I phoned Hugh Jones, a friend and mentor who part-owned Airwork. I told him what had happened.

'John, don't think of it as a disaster,' he told me. 'Think of it as an opportunity.'

That's the kind of guy Hugh is. He's a wheeler-dealer, and he has an extraordinary ability to turn reversals in fortune to his advantage.

But, despite his inspirational words, I couldn't see any way forward, and I phoned him back after a couple of days and sleepless nights to tell him so. He told me to drive up to Auckland to see him.

'I just can't see any opportunity in this situation,' I told Hugh as I sat in his office.

'Take a look out there,' he said. I looked out the window at a Squirrel, identical to mine. 'You can fly home for the price of what it costs to fuel up if you promise my company gets the rebuild of your machine.'

There was a ray of hope, after all.

I phoned the insurance assessor and told him that Airwork had to get the job.

'Well, they'll get the job if they put in a competitive tender,' he replied.

Hugh got to work on it, and probably because he had quite a few second-hand spares in stock from other rebuilds, he managed to submit the winning tender. The deal was on! I was back in the air.

But as days went by, it looked like we might be in trouble, after all. Airwork wanted progress payments on the work of rebuilding the chopper, and my insurers weren't paying out.

Upon enquiry, it turned out that there's this weird rule in the insurance industry that unless you've paid your total premium, they won't pay out on a claim. I had crashed my chopper partway through the insurance year.

'Can't you just deduct the premium from the payout?' I asked.

No, they said. They couldn't. I tried to get my insurance broker to resolve the matter for me, but he was very slow to return my calls.

Hugh was pissed off about this, as he wanted progress payments to get the parts on their way to start the rebuild.

'Let's go and visit this guy,' he said. So we went to the Auckland offices of Minet Mathers, my insurance broker, who in those days was a represented by a man who rejoiced in the name of Robin Hood. We asked his receptionist if he was available.

'No, I'm sorry,' she said. 'Mr Hood is out. I can tell him you called and get him to contact you, if you like.'

'No, we'll wait, thanks,' said Hugh, with a winning smile. We took seats in the waiting room. Time went by.

'I don't know when Mr Hood will be back,' the receptionist said after half an hour had gone. 'Are you sure you don't want me to get him to call you when he gets in?'

'No, thanks,' Hugh said. 'We've got nothing better to do. We'll just sit here till he turns up.'

All parties knew that Mr Hood was in his office within earshot. Hugh and I talked loudly about what we thought of the situation, and made all the obvious jokes about robbing from the rich, giving to the poor and so on. Eventually my less-than-pleased-looking broker emerged. He showed us curtly into a meeting room where a heated debate ensued.

To his credit, Mr Hood acted just like the Robin Hood of Sherwood Forest and phoned the parent company, London insurance broker Minets. Their cryptic reply, when it came, was 'There is no taste in nothing', which (as any underwriter knows) means you have to have paid your premiums in full before any claim will be paid out. So Minet Mathers in New Zealand paid the premium, simply so that Minets in London would make a progress payment, from which the premium was promptly deducted. Our insurance premiums went through the roof after that, and now we had a debt to Hugh to go with all our other debts. But it hardly mattered. We were back, thanks to Hugh Jones. Business is like a horribly real game of snakes and ladders. We'd been on a bloody great snake, but he put us back on a ladder.

***

Who knows how many lives Hugh inadvertently saved by getting us back in business. We were in the search and rescue and air ambulance business. Throughout the whole changeover from Helicopters New Zealand to private ownership, we had continued to perform search and rescue services, which we were basically funding out of our own pockets. Shortly after we took over the company, we were doing up to three search and rescue call-outs a week.

We had been starting to attract a bit of attention. Some cases seemed to pluck at the heartstrings of the public more than others: Dr Pearl's awful death in the Huka Falls was one example, and in the wake of that episode Peter Minson and Sue Hawkins, the senior people at the local Taupo radio station, ran an appeal to raise funds for new and better rescue gear. The idea was to pull in $10,000, but we did better than that. The final sum was $16,000, which enabled us to buy a purpose-built stretcher and other bits and pieces, such as a battery-powered device that gave a constant readout of the amount of oxygen in the blood via a probe that was attached to the patient's finger and a bag that allowed much of the emergency medical supplies to be carried in one pack.

Another high-profile case we took on was in August 1980, when we were called in to help look for a pair of young girls who had gone missing in the Urewera Ranges. Salina and Noti Apirana had gone into the bush from Ruatahuna to work a possum trap line. They were lightly clad, and when they hadn't returned by nightfall, a search was mounted. The weather closed in. It was cold, with torrential rain, and days passed with no sign of the pair. Around 100 searchers were pushing their way through the bush on foot, and it was looking pretty grim.

Then, against the odds as it seemed, we got word to stand by to help extract someone from the headwaters of the Ruatahuna Stream. This was kind of inconvenient, as the Lama was undergoing maintenance and was in bits on the floor of the hangar. We put it back together again as quickly as we could, and sure enough, the call came through to fly to Murupara. I was reasonably sure it would be a body recovery mission, but word reached us that they had found one of the sisters alive. Thinking it likely we'd need a bit of manpower to help with loading up a patient from steep ground, we picked up a couple of firemen and then flew to Ruatahuna to collect Charlie Whiting, the search coordinator. Then we headed for the hills.

The combination of adverse weather and the steep, heavily forested terrain made this one of the more difficult flying assignments I had ever undertaken. Once we located the area of more or less open ground in the streambed where the rescuers were huddled around the blanket-swathed form of 20-year-old Salina Apirana, I got us as close to the ground as I could. There was no chance of landing. As I hovered there, Salina was secured in our fancy new stretcher and was then lifted overhead by the gang on the ground and passed into the helicopter cabin, where Dean and the firemen helped to settle her in position.

But it wasn't over yet. We touched at Ruatahuna so that the police could have a word with Salina to see if she knew where her sister might be, but it was clear she needed to get to hospital as soon as possible. She was hypothermic and there was a lot of concern about her feet, which were waxy with frostbite. So we took off into the gloom again and headed for Rotorua. A few miles short, I worked out we didn't stand a chance of getting into the Rotorua basin, so we diverted to Taupo. Again, we were thwarted. Whakatane looked like our best option, so we made a dash for the coast—only to find it was no better there. Ironically, the clearest area we could locate was back in the Urewera, where Murupara had opened up. After the best part of an hour stuffing about in the air, we set down at Murupara where Salina was loaded in an ambulance for the drive through to Rotorua.

That still left poor 19-year-old Noti unaccounted for out in the bush, but the positive outcome for her sister renewed hope that the younger girl, too, might be found alive. Sure enough, the following morning we heard that an RNZAF Iroquois was inbound with Noti aboard, alive but obviously very much the worse for wear after nearly a week in the frigid bush. Once she had been checked out, I flew her to Rotorua Hospital, before returning to help the Iroquois lift the cold, tired but very satisfied search teams out of the bush.

The press were all over us, loading all the credit for the sisters' survival onto us—never mind all those poor blokes who had slogged through some of the most inhospitable terrain in the North Island in atrocious weather. I tried to put the record straight on that, but it was no use. It's apparently much easier to single someone out for praise than to lavish it on a hard-working team.

***

Still the rescues kept on coming. I was obliged to fly in marginal conditions again when a bunch of 10 Otago University students were caught out in a blizzard on Tongariro. A pair of them managed to walk out to raise the alarm, but the rest were forced to build an igloo for shelter, and we had to take them off piecemeal as the weather allowed. While the live deer recovery boys worked hard to capture animals—a live hind was selling for $1000 straight from the helicopter at the time—we were kept busy recovering them when they pushed the margins too far. And the unforgiving Central Plateau roads saw to it that there was always a steady stream of maimed and injured from traffic accidents, too.

In mid-1982, the Taupo Moana branch of Rotary invited me to a dinner where I was presented with a commendation for my services to search and rescue. Lots of the people I'd worked with—police, medical staff and firemen—attended and spoke about the rescues I'd been involved with. It was all pretty awkward. To listen to them, you'd have been forgiven for thinking I was a one-man rescue team doing it all myself. And, as if that wasn't enough, two years later, around the time we took over the company, I got a letter through the mail on a better class of stationery than you're used to getting when all you receive is bills. It had the letters O.H.M.S. embossed across the top, and it took me a couple of readings to convince myself it wasn't a joke. I suspected a number of friends and family must have known about this and had not informed me. So I hid the letter and kept it confidential until the day before New Year's Day, when the press, who had been under an embargo, rang Trish to ask her whether they could speak to me about my honour. Her Majesty wanted to make me a Member of the British Empire for my services to search and rescue!

'How long have you known about this?' Trish asked.

'A few months,' I shrugged. 'How long have you known about it?'

'I had no idea,' she replied.

It turned out some prominent members of the community had nominated me for the award. It was humbling and, quite frankly, a bit ridiculous.

Still, our steadily rising public profile was proving useful in raising funds for rescue equipment, even if we were still doing SAR work for the sheer pleasure of helping a person in need. In 1984, however, events conspired to give us the opportunity to put the whole rescue side of things on something resembling a professional footing.

***

Late in the morning of Wednesday, 17 October 1984, the phone rang. It was Terry Knight, calling from the Wellington Rescue Coordination Centre.

'Funnell, we've got a job for you. There's a light plane took off from Turangi this morning on the way to Palmerston North. Call sign Echo Foxtrot Zulu. He's 70 minutes overdue. Reckon you could go have a look for him down the Desert Road?'

The wind was buffeting the house, and there was a low cloud cover down over the lake. It wasn't a day that exactly filled you with the desire to jump in a machine and go flying.

'Okay,' I said. 'Do we know when he was last heard from?'

Paul Taylor filed his flight plan a little after 0832 that morning. He was flying his little four-seater Grumman AA-5A Cheetah under charter from Turangi to Wellington via Palmerston North with two passengers aboard. Hugo Schut, 53 years old, was the managing director of Philips Electrical. Janeane Robinson, who was 28, was the company's promotions officer. Both lived in Wellington where the company's head office was located, and had flown up in the same plane the day before to attend a conference in Turangi.

Taylor himself was quite young, only 31. I didn't know him, but it turned out he was well thought of by his colleagues in the Wellington Aero Club as a good pilot, and—ominously, in his present situation—was known for being pretty meticulous in his communications. His announcement to air traffic control that he was airborne and expecting to be abeam Palmerston North 45 minutes later was his last communication.

I asked whether we could expect there to be an emergency locator beacon (ELB) signal if the plane had come down.

'There's some confusion over that,' Terry told me. 'I'll get back to you.'

By the time I was climbing into the Squirrel to have a look along the planned route, I had been told that there was no ELB. The plane was normally fitted with one, but it had been removed for servicing and wasn't ready by the time the plane left Wellington.

Here we go, I thought. Needles and haystacks.

***

The usual route from Turangi to Palmerston North was to get aloft and then to follow the Desert Road through to Waiouru and Taihape and on down State Highway 1, veering east by south from around Hunterville to Palmerston North. There was good reason, however, to believe Taylor could have deviated from this plan, especially with the visibility coming and going.

I flew along the road to try to remain below the cloud ceiling, but even so, the road—a ribbon of black tarmac in the duncoloured tussock of the Rangipo Desert—was difficult to see at times. The wind was ferocious—it was gusting as high as 70 knots (125 kilometres per hour)—and so was the turbulence. Even the Squirrel bounced about in the so-called 'rotor belt' in the lee of Mount Ruapehu, one lurch violent enough to lift me off my seat. I could imagine how hard it would be to handle a light, 150-horsepower fixed-wing aircraft in these conditions.

It was pretty plain we had a search and rescue operation to conduct, and with the forecast indicating little let-up in the adverse weather, it was going to be a tough one.

In order to try to narrow down the search area, the first job was to gather intelligence—to mobilise the police, and everyone's networks of friends and relatives in the area to pick up the phone. We went door-knocking along the route and in adjacent areas to find out if anyone had seen or heard anything of the missing plane. Trouble is, the route lay along the Desert Road, and it isn't called the Desert Road for nothing. Once you get past Rangipo Prison, there isn't really anything in the way of human habitation. Still, we did what we could. We had firm sightings of the plane shortly after its take-off from Turangi. We soon began to get reports from people who had seen a small aircraft over the Central Plateau, or who had heard one. Some of these indicated the plane may have been off the expected route and flying down the west side of the volcanic massif of Ruapehu and Tongariro. At about 0900, someone who had been travelling on State Highway 47 near Lake Rotoaira (west and a little south of Turangi) reported seeing a single-engined plane heading in the direction of the Chateau on the slopes of Mount Ruapehu. Several people in the tiny township of Waikune (just south of National Park and due west of Mount Tongariro) saw a small plane pass over at about 0915. Someone in Horopito heard a light plane pass overhead around the same time, or soon afterward: the cloud was too thick for a visual sighting. The police received a report from someone at Rangataua, pretty much due south of Ruapehu, which very closely matched the description of the missing plane. At that point, no description had been issued. And nearby, from the workers at the Karioi weighbridge belonging to a timber mill, came an electrifying account of 'a loud explosion, followed by a thump and a high-revving noise'.

By now, we had eyes in the air. Within six hours of commencing the search, we had a number of helicopters and fixed-wing aircraft up and looking, and the Wellington Rescue Coordination Centre (RCC) had assumed control. A 100-man ground search team had been placed at their disposal by the army, and more and more aircraft and observers came forward to assist. On day two of the search the job of coordinating this effort fell to me and to the Taupo-based Flight Services Officer, Roy Carmichael. Between Roy and me, we had quite a bit of search and rescue experience. We decided that the weight of the evidence suggested the plane had taken a course down the western side of the volcanoes, and that it had made it at least as far south as Ohakune. We decided to base ourselves in the little town of Raetihi, just under 10 kilometres due west of Ohakune. Louisa Patterson, a helicopter pilot employed by Wanganui Aero Work, was based there, and she made her hangar available for a headquarters.

By mid-morning on Thursday, 18 October, we were well established there, with heaters going and phones ringing and a blown-up map of the Central Plateau, photographs of Echo Foxtrot Zulu and pins representing the various sightings on a display board on the wall. People were coming and going in droves: pilots would come in for a briefing, and we'd tell them what the latest intelligence was indicating, the weather and visibility conditions they could expect, and we'd assign them a sector to search. Many of the volunteer observers weren't that used to flying, so plenty who went out all bright-eyed and bushy-tailed came back for a debrief looking pretty sick and shaken by the experience of freezing their tails off in choppers easing through the sleety showers with the doors off, or being bounced around in a fixed-wing pitching and yawing in the turbulent wind. Most stuck with it, though, even as the hours went by and the conviction grew that this was more of a search than a rescue.

At some point someone nodded toward a middle-aged, careworn man who was standing and watching the goings-on.

'His dad. The pilot's.'

***

The locals of Raetihi were amazing. The pub put up everyone—pilots, observers and coordinators—and an army of local women kept us fed and watered and served sandwiches, soup, hot drinks and gingernuts between meals. Farmers would drift in with shoeboxes of hot scones: everyone who couldn't help with the actual searching seemed to find a way to assist, as Kiwi communities do in these situations. By now we had 11 helicopters and 10 fixed-wing craft aloft and looking for the missing plane. To facilitate communications between them, we had Arthur Whitehead in a Cessna 206 belonging to Air Charter Taupo scribing wide circles at 10,000 feet, listening and relaying instructions from the search aircraft at lower levels.

Still little snippets of information kept coming in. Niko Stephens, who farmed a block on State Highway 49 near Tangiwai, had heard a plane pass over two or three times on Wednesday morning, as though it was circling. Was it lost? he wondered. We wondered the same thing, but we were also aware that one of the problems of SAR work is that the search itself can contaminate the intelligence you're receiving. People desperate to help think back and remember hearing a plane: without being able to fix the timing with any precision, it's just as likely they heard an aircraft involved in the search as the target. But a woman driving near Rangitaua was able to give a very detailed description of a blue-and-white, single-engined plane flying in the direction of Ohakune; she didn't see the full call-sign, but she noticed the final digit was the numeral 2. It wouldn't take much to mistake the Z in EFZ for a 2. This sounded promising, and it reassured us we were looking in the right area.

Meanwhile, we learned that a prison officer at Rangipo reckoned he had seen or heard a plane. Trouble was he had gone off-duty when we tried to chase him up and no one seemed to be able to find him. This sighting was much closer to the plane's expected route. And we heard from a man who had been scrubcutting near the corner of the Desert Road and Paradise Valley Road. He reported seeing a single-engined plane fly over, heading in the direction of the Moawhango Dam. This sighting was south and a little east of Turangi and well to the east of where we were concentrating the search. And looking at the map, and the high and forbidding terrain that lay in the direction indicated, you couldn't believe anyone would fly up there. Besides, the Kaimanawas were clagged in with cloud, making that area impossible to search.

Still, you have to keep an open mind. Sometimes, it will surprise you how wide open you have to keep it. At one point on Thursday afternoon, George Taylor, the missing pilot's father, approached me. He looked haggard, and he also looked uncertain about what he was going to ask me.

'Mr Funnell, this might seem like a funny question,' he began, and coughed.

I waited.

'Do you believe in the supernatural? Y' know. Clairvoyance?'

I kept my expression as neutral as I could, and shrugged.

'I've never really thought about it,' I replied. 'But if anyone thinks they have information that might be useful...'

He looked encouraged.

'My brother Lionel got a call from a woman named Francie Williams. She claims she's a psychic, and she says she knows exactly where the plane is. She mentioned a canal, and something that looked like a fire tower. Does that ring any bells with you?'

It didn't, but this didn't mean I was going to dismiss the information out of hand.

As a matter of fact, I'd had a little bit to do with a psychic the previous year, when I was coordinating a search for a plane missing in the Southern Alps. Roy Turner was flying his wife and two preschool kids about in his Cessna 172 when they disappeared in the Fox Glacier area. Two weeks of aerial searching had failed to find them when the family decided to contact June Donaghie, an Auckland-based clairvoyant. She claimed to be able to sense the location of the plane by the hot spot she could detect on a topo map. She also claimed that Roy was personally feeding her the information, and she used a salty turn of phrase that was distinctively Roy's and quite obviously not hers. She also seemed to know a lot about the airplane that hadn't been publicised. The coordinates she gave me over the telephone were eerily plausible—so plausible that I thought it might be worth getting her involved in the search.

When she arrived at Fox Glacier, I arranged for her to go up in one of the helicopters with Colin Tuck, and Tucky reckoned she got excited right when they were over the spot she had marked on the map. It was a brutal bit of terrain, completely inaccessible and subject to avalanches. No trace of the plane was visible, but it was quite possible it was there. Like me, Colin would probably have dismissed the whole thing had she not pointed out spots on their route back to base that she felt had been the sites of other tragedies. Colin was aghast. In each instance, she was right.

So I gave George a map.

'Take this to Francie and see if you can get her to mark the spot on the map.'

He took the map and went away.

***

The search was by now reckoned to be the biggest in New Zealand's history at that time (although there have been bigger and costlier since, such as the search for Michael Erceg's missing helicopter in 2005). It was costing an estimated $15,000 a day, and, inevitably, rumblings began to be heard from Wellington. They never cancel searches: only suspend them. As Labour Weekend loomed, the weather continued poor and helicopter after helicopter touched down with the searchers drawing a blank, suspension looked to be a very real prospect.

On Friday, I was approached by another character I'd noticed hovering on the periphery of things. He was a big bloke, and something about his clothes indicated he was a city slicker rather than a local. He was dressed casual, but not in the standard-issue Swanndri bushshirt, black singlet, musterer's hat and gumboots that prevailed. He'd asked the odd question every now and again, but mostly he'd seemed happy to stay in the background watching everyone at work.

'Mr Funnell, John Cooper,' he introduced himself. 'I'm the group manager of Philips New Zealand.'

We shook hands. John told me he had heard rumours that Wellington was going to suspend the search.

'The parent company in Holland are keen to carry on the search at company expense,' he explained. 'Are you and Mr Carmichael prepared to carry on in your roles?'

I was still in the same clothes I'd been wearing when I left Taupo on Wednesday afternoon. I could use a change of clothing and a good night's sleep.

'Absolutely,' I said.

***

The privately funded search kicked off with a briefing we gave to all of the pilots and to the media on Saturday, 20 October. We explained that we were at a bit of a loss, and that the search area had been widened outside the areas of probability to include areas of possibility. A combination of the weather and of the terrain we were searching—the featureless expanse of the Rangipo Desert was flanked by steep, bush-clad country, more often than not gouged by watercourses—meant that we couldn't say with any certainty that the plane lay outside our initial search area, but that certainty was growing. We were now looking well to the south of the Central Plateau in the direction of Taihape, on the basis that had Echo Foxtrot Zulu suffered a communication failure, it may have flown far further along its flight path than we had presumed. Similarly, if Paul Taylor had been battling severe turbulence, he might have been prevented by the sheer physical effort required to keep things straight and level from operating his radio. Whatever the case, although the needle seemed to be the same size, we were facing the need to look through a lot more hay.

Many of the faces turned attentively towards me during the briefing were showing the signs of the toll the search was taking. The surest indication that it was hard yakker was that the helicopter boys were so well-behaved. They were a pretty colourful lot in those days, mostly young and with a broad streak of the larrikin about them. But even though they were being put up, all expenses paid, in a pub, there wasn't a man jack who didn't have a quick bite to eat at the end of the day, washed down with a beer or perhaps two, a chat with the other pilots and then hit the sack early, ready to do it all again at first light the following day.

By Sunday afternoon, with no sign of the plane turning up, we were really starting to scratch our heads. A Philips company car showed up with Renée Schut—Hugo's wife—and their three daughters, the youngest 16 and the oldest 24. I gave them a briefing and was forced to admit that we were flummoxed. Renée gave me a sad smile that was not without sympathy. She thanked me for all that we were doing.

I learned there was a message for me on my home answer phone from Francie Williams. She was adamant that we were searching on the wrong side of the mountain. She claimed she had information from a young yachtie who had drowned off the Northland coast in 1983 and who had made it his mission in the afterlife to assist those afflicted by tragedy. He had told her that they should be closely searching the area in the vicinity of the Pillars of Hercules.

This meant nothing to me. I was beginning to get frustrated with Francie, whom I felt was distracting everyone from their work. But George Taylor reappeared, and spread out the map I had given him.

'Francie reckons the plane is here,' he said, pointing to where she had marked a neat cross in red biro. It was at a site on the western slopes of the Kaimanawa Ranges. The topo map indicated it was steep country, and probably in thick bush.

If Paul Taylor had got up there, we'd have to believe he steered hard left from his expected route. I couldn't see any reason why a competent pilot would head into tiger country like that, especially in poor visibility.

'Interesting,' I told George. 'That area's been in cloud since day one. But as soon as the weather clears, we'll have a look in there.'

George nodded.

***

That evening, even the pilots were beginning to mutter that any further searching was futile. We've looked everywhere, was the general feeling. On Monday morning, I had a discussion with John Cooper in Raetihi in which I told him

I thought it was time we scaled back the search and reviewed all the information we'd received, in case we'd overlooked something vital.

'I agree with you,' John said. 'But I've got head office in Holland on my back. They want results.'

He suggested that before the pilots were dispersed, they should each be allowed to spend an hour to pursue their own hunches. Philips would pay. I couldn't see any harm in this plan, so we briefed the pilots accordingly.

But as the rain rattled on the roof and the light began to fade on Monday afternoon, it was plain that even this initiative had failed to turn up anything. We began to pack up, and I headed home for a much-needed break, intending to take time to review all the intelligence in case we had missed something. Experience tells us the information or indicators are inevitably present. The problem is the one vital piece is amongst thousands of other very similar and equally convincing reports.

John Cooper agreed to suspend our efforts for a day.

'Funnell,' he said. 'The day after tomorrow there's going to be a board meeting at Philips here in Wellington where the board are going to ask me how it is that we've spent over $200,000 and have nothing to show for it. And you know what? You're going to come down to help me explain it all to them.'

***

Strictly according to Murphy's Law, the weather on Tuesday morning was glorious. The phone rang. It was a man from Waipukurau named Bay de Latour, who was a friend of Janeane Robinson's. He had a Cessna 172, he told me, and it was such a nice day and besides, he wanted to help. The Robinson family wanted him to have a look for Echo Foxtrot Zulu. Was there anywhere I thought he should look?

The sky over the Kaimanawas was sparkling blue.

'Well,' I told Bay, 'there's an area just north of Moawhango...' and gave him the grid reference.

I happened to be near my VHR radio set at home early in the afternoon of Tuesday, 23 October when a call came across. It was Bay.

'Hey, John,' he said. 'You won't believe this. We're up in the Kaimanawas and we've found a burned patch up here.'

At about the same time, Richmond Harding, a pilot from Wanganui Aero Work, came on. Richmond had known about our interest in the ranges and when he'd seen conditions were right, he and his mate Chris Wolff had taken it upon themselves to have a look, too.

'Wolffy reckons he can see something that looks like part of a plane. You want to send someone up here to check it out?'

All kinds of emotions went through me. There was cautious scepticism—this spot still lay well outside the areas of probability—but I could already feel relief pricking at the backs of my eyes, a kind of weariness settling over me and the growing sense that the end of the last week's massive effort was in sight.

I contacted Toby Clark, a fellow chopper pilot, a real ace with lots of experience in venison recovery in difficult terrain and a good knowledge of the area in question. Toby and his mate Jeff Prince agreed to fly their Hughes 500 up and have a look.

An hour later, Toby called.

'Yep,' he said. 'I've just long-lined Princey in there. It's your plane. It's burned out, and there are three bodies in there. Not too pretty, Princey reckons.'

I got off the radio and then called the police and RCC. Then I phoned John Cooper.

'I've got some news for you,' I told him.

'Good or bad?' he asked.

'Well,' I replied, 'sort of both. We've found your aircraft, but there are no survivors.'

There was a long silence at the other end of the phone.

***

The following morning, I was taxiing the Air Charter Cessna 206 I had borrowed from Arthur and Christine toward the runway at Taupo airport when I saw the figure of Jeff Prince walking toward me on the tarmac. He looked pretty bedraggled—unshaven, in jeans and T-shirt and with a haunted look on his face.

'How are you?' I asked him.

'I've just been helping with the recovery of the bodies from the Philips plane crash,' he said, which explained the sombre look on his face. 'Where are you off to?'

'Heading to Wellington,' I told him. 'Philips have asked me down. The Dutchmen want to know how we've dusted off all that money. Want to come with me?'

Jeff looked down dubiously at his clothes and rubbed a hand over his unshaven jowl, but then seemed to make up his mind.

'Okay. Don't mind if I do.'

He jumped in beside me, and we flew down to Wellington. We followed the Desert Road, keeping the line of power pylons on our right-hand side. South of Rangipo, both of us glanced involuntarily left toward the high, brooding country where the plane and its hapless pilot and passengers met their end. Jeff told me that he and a couple of policemen had been dropped at the site. One of the bodies was lying a few dozen metres clear down the slope. The other two were in the mangled and charred remains of the plane. It was clear that it had gone in with some force. After telling me the details, Jeff was pretty quiet for the rest of the one-hour, 20-minute-long flight to Wellington.

***

John Cooper met us, and his handshake was the handshake of a grateful man. He drove us into town to the tall office building on Lambton Quay where Philips NZ had their headquarters. We rode the lift up to the reception lobby, and felt as out of place in our casual (in Jeff's case) backcountry clothes as we looked amongst the suits and ties, but no one seemed to care. On the contrary: we were greeted like some kind of conquering heroes. There was a very attractive young lady on the reception desk, and she stared at us with open admiration, especially the rugged Jeff. All the same, Jeff felt uncomfortable.

'Jeez, I haven't shaved for a couple of weeks,' he muttered.

John Cooper waved a hand at the framed advertisements in the reception area, many of which featured Philips shavers.

'No problem there, I reckon,' he said, and so it proved. The receptionist quickly found a shaver for Princey, and he emerged from the dunnies looking pretty smooth.

We made a brief appearance before the board. The distinguished Dutch gentlemen and others sitting around the table expressed their gratitude to us, and contrary to the popular national stereotype, didn't seem to care too much about the money, even when I apologised for how much we'd spent. After all, we had got a result of kinds.

Naturally, they wanted to know what had happened. I told them that what I imagined was going on that morning was that a young commercial pilot, anxious to please his client and scenting the possibility of further work if he could manage it, made a bad call. The plane was barely suitable for the flight, and distinctly unsuitable to make that flight in those weather conditions. Paul Taylor's instincts and training will have told him that the prudent call was to cancel. But he chose to take the chance.

Once aloft, with a reasonably high cloud ceiling, he was able to maintain some sort of visual fix on the ground. He would have been keeping the line of pylons on his right. He probably lost sight of it momentarily as a clot of thicker cloud blew through, and then dimly perceived the line of pylons passing beneath him. He would have corrected to bring them back onto his right wingtip and breathed a sigh of relief—unaware that he was now following the feeder line that intersects the Desert Road main trunk and that originates from the Rangipo Power Station high up in the Kaimanawas. The first inkling he would have had of trouble would have been the wall of the ranges emerging from the gloom ahead and on both sides. It's the stuff of aviation nightmares. What he seems to have done then is to have given her full throttle and tried to outclimb the hills. Then, in the lingo, it would have been CFIT (controlled flight into terrain). It would have been a high-speed collision, and the only mercy for the relatives of the deceased is that death will have been instant. The plane then caught fire and burned out nearly an acre of bush. In better weather, we would have found it far sooner.

I told the Philips board and the media that had the plane been carrying its emergency locator beacon and had it worked and not burnt, it would have taken less than two hours to find it. All that effort, all that expense, the whole awful rollercoaster of hope and dread that the families suffered—there would have been no need for any of it.

After the meeting, the receptionist lifted up a sliding panel to reveal an impressively stocked bar. We were invited to help ourselves. For Jeff and I, who had been hard at it for two weeks, no further encouragement was required. Jeff, in particular, had the memory of a few sights and smells to erase.

'Rum and Coke,' he told our hostess, holding up his thumb and forefinger in a pinching gesture. 'Rum, with just a little dash of Coke.'

She obliged! Jeff had a fair tolerance for rum, the kind you acquire with practice. But what, with the fatigue and the quality of the pour, he was soon under the influence. I wasn't far behind despite my much more moderate consumption rate. John Cooper took us in hand, insisting that we stay the night with him. We hadn't thought that far ahead, so we were happy to accept. And it turned out his daughter was having a birthday party with a few guests around, and we had been billed as some kind of star attraction, the intrepid heroes of airborne search and rescue. Well, as entertainment, we were a bloody disappointment. By eight o'clock, when most of the guests had arrived, Jeff and I were sound asleep.

***

The tangible expression of the gratitude of the Philips organisation wasn't confined to generosity with the corporate liquor cabinet. A couple of weeks after we had returned to Taupo and life had more or less returned to normal, John Cooper phoned me. He told me that everyone watching the search had been impressed with what it is helicopter search and rescue services can achieve. They wanted to show their gratitude to the people of the Central Plateau by supporting our operation, and to both equip and fund us so that we could perform at the kind of level Peter Button was doing in the Wellington region. Peter was a very capable and experienced chopper pilot, and had been doing heli-rescue work since the early 1980s. Thanks to a generous sponsorship deal, his operation was very well resourced. To achieve this, a trust would be set up, to be called the Philips Search and Rescue Trust. John Cooper would chair it, he told me, and he invited me to think of the names of local people suitable to share the work. I had no problem doing this.

A couple of months later, by mid-December 1984, the trust was up and running, with John as chair, and myself, Taupo surgeon Terry Burcher and Tongariro National Park chief ranger Bruce Jeffries as trustees. The Squirrel was signwritten with the name of the Philips Search and Rescue Trust and launched to the media in March 1985, and it wasn't just the looks that had changed. We had a whole new suite of medical and search equipment aboard, courtesy of Philips—purpose-built stretchers, life-support gear, a new and better radio, radio direction finders, a million candlepower 'Nitesun' lamp and a winch—and we also had annual operational funding to the tune of $30,000, enough for around 30 flights. There's little doubt that the thick grey cloud that settled over the Central Plateau in October had tragic consequences for Paul Taylor, Hugo Schut, Janeane Robinson and their families; but for countless other people since, there has been a silver lining.

There was another sequel to the disappearance of Echo Foxtrot Zulu, too. I was obliged to somewhat alter my world view. When I showed them on the map where the plane had gone in, Renée Schut was astonished. She told me that when she and her daughters had driven past the vicinity, all of them had experienced the strong conviction that Hugo was trying to communicate with them. When they got to journey's end, they all reported the same feeling. And when I was talking to George Taylor, he told me that when he drove past that spot, he was so convinced that Paul was close by that he actually stopped his car and got out and called: 'Where are you?' into the mist.

And then, of course, there was Francie. On the map, close to the crash site, there is a spot where the Tongariro River plunges through a narrow gorge flanked by a dramatic rock formation known as the Pillars of Hercules. One of these is so tall and regular that it could easily be mistaken for something man-made—a fire tower, for example.

Upon my return from Wellington, I phoned Francie.

'Francie, you know where you reckon the plane crashed? You put a mark on the map to show us?'

'Yes?'

'Well, sorry to say, you were wrong.'

'Oh!'

'Yep. You were thirty metres out.'

The Taupo-based Philips helicopter is readied for an infra-red search for a missing person.

CHAPTER SIX

TURBULENT TIMES

No sooner had we got the Philips Search and Rescue Trust up and running and the Squirrel fitted out with all its new gear than we suffered a fairly significant setback. I was up in Auckland in June 1985, training for my Instrument Flight Rules rating on the aircraft, which would allow me to fly by instruments and therefore in a greater range of conditions—at night and in cloud—when I got a call to say that John Fogden had crashed the machine. It was a write-off.

Foggy had been working up on the skifields on Tuesday, 4 June, and he made the decision to fly back to Taupo even though the weather was closing in. It was an understandable decision. We tended to find that bits and pieces disappeared off the machine if we left it tied down up on the mountain. But as he neared Taupo, it was just about dark and a thick fog was setting in. Foggy radioed Roy Carmichael in the tower at the airport to get assistance in making an approach. He was hoping he could use the runway lights to give him a visual reference as he got close to the ground but, unfortunately, as he was making a turn to line up the end of the runway, the rotors clipped a tree and the whole thing came down. The rotors were bent beyond repair, and the fuselage wasn't much better.

I was devastated. We had been working hard to convince people that helicopter aviation was safe. The machine had just been fitted out with all the SAR bells and whistles. And to make it worse still, the Squirrel was painted in the Philips Company livery.

Foggy was contrite when I talked to him on the phone.

'Ah, well,' I told him. 'We can't undo what's done. You okay?'

He was lucky not to have been injured, or worse. We managed to rent a replacement machine from my good friend Grant Biel in Auckland and much of the navigation and rescue gear, which was undamaged, was switched across. We got out of it quite lightly, all told.

***

There's an old Chinese curse: 'may you live in interesting times'. New Zealand was cursed in the 1980s, it seems. It was one of the more turbulent decades, socially and politically speaking, in the second half of the twentieth century. Even humble helicopter pilots couldn't keep out of the way of it all, all of the time.

In the early afternoon of Saturday, 25 July 1981, back when we were still running the

Lama for Helicopters New Zealand, I got a call over the radio from Roy Carmichael at Taupo flight services asking if I could assist in tailing a light aircraft that was heading away from Taupo into the Waikato. They got me to pick up Constable David Jensen, who filled me in as we got airborne.

At about 1.30pm, a well-dressed, elderly bloke had strolled into the offices at Lakeland Aviation and asked to use the toilet. The staff were happy to let him, but a minute or two after he had disappeared through the back door, they heard the engine of a Cessna 172 that was being readied for a scenic flight start up out on the tarmac. They ran out to see what was going on, but the man—his name turned out to be Pat Macquarrie—gave her full throttle and started taxiing out onto the runway. There was nothing anyone could do to stop him.

Graeme Stratton, a Lakeland pilot, was just on his approach to the aerodrome when Macquarrie took off. Graeme landed his Cessna 185, unloaded the three hunters he had aboard, and quickly turned around and took off again to give chase. That's when we were called.

A hundred or so nautical miles away, the second match of the Springbok tour was scheduled to kick off in just under an hour's time. The tour had bitterly divided New Zealand society: in the face of strong opposition, the Prime Minister of the day, Robert Muldoon, had pledged his government's support for the New

Zealand Rugby Union to invite the Springboks to play a full tour. South Africa at the time was governed by a regime that practised apartheid, and had been ostracised by most of the international community. When it became clear that the tour would go ahead, a very highly organised and determined protest movement got under way. They vowed to use any means at their disposal to disrupt the games, and the police began tooling up with new gear and tactics. It looked like it was going to get ugly.

After the protest movement mounted a concerted effort to stop the first game at Gisborne on 22 July, but were prevented by the police, a rumour went around that a plane would be used to stop the next game, perhaps by dropping stuff on the field, or even being deliberately crashed into the crowd. Pat Macquarrie himself had apparently let it slip while drinking at the Waiuku Rugby Club that he had terminal cancer and thought he might try to go out in a blaze of glory and take the Springboks with him. On the Friday, the police issued a request to all aircraft owners to make sure their craft were secure, and they began trying to keep tabs on Macquarrie's movements.

The 172 scribed a few wide circles over Wairakei, which was surrounded by cloud. Then, as the cloud cleared, it headed off north-west, following State Highway 1. Graeme Stratton managed to manoeuvre his plane alongside and tried to get the pilot's attention, gesturing back towards Taupo. He radioed to say that Macquarrie had glanced at him, but he'd made no acknowledgement or move to change course. Similarly, flight services were having no luck raising him on the radio. It was beginning to look ominous.

We caught up with Stratton and Macquarrie as we flew over Tokoroa. The plane that Macquarrie was flying was capable of doing a little over 100 knots comfortably. I could push the Lama to 110 knots if I had to. I could catch him okay, but the question was: what then?

Meanwhile, we learned that a bunch of protesters had broken through the perimeter fence around Rugby Park and had formed a tight knot on the playing field. Tempers were running high, we were told, with the police obliged to switch from trying to prevent the protesters from gaining entry to trying to protect them from the irate rugby supporters packing the stands. We were constantly relaying our course and speed to police headquarters, and it was plain that anxiety levels were rising.

'How many flying minutes from Hamilton?' they kept asking.

We flew over Tirau, and whereas I had been expecting the fugitive pilot to veer north by west toward Cambridge, he pressed on toward Matamata. There were still patches of low cloud, and I began to suspect he was uncertain of his position. As we neared Morrinsville, the same call came over the radio.

'How many flying minutes from Hamilton?'

'Ten,' I replied.

There was a long pause before the radio came back to life.

'Match abandoned. Match abandoned.'

Almost immediately, the plane ahead of us veered to line up the racecourse at Morrinsville.

'Looks like he's going to land,' I said.

'I'm getting low on fuel,' Graeme reported. 'I'm going to have to turn back to Taupo soon.'

The 172 performed a copybook landing on the paddock in the middle of the racetrack, and I set down close behind him as a Ministry of Transport plane circled overhead.

'Get over there and open the left-hand door,' I told Dave Jensen. 'Grab the bugger before he does anything.'

We were thinking explosives, or firearms, or the possibility that the plane might try to take off again.

I ripped off my headset and got my own door open as Dave rushed the plane. He made a grab for the pilot's door handle, but his feet went out from under him on the slippery turf. He landed on his back and slid past the nose and the propeller, which was just winding down.

I was close behind, and got the right-hand door of the plane open. Macquarrie shot me a panicked look and started fumbling in his jacket pocket.

'Get your hands up!' I yelled, and he meekly complied.

By now, Dave Jensen had got back to his feet and he tore open the pilot's side door. He sprang Macquarrie's seatbelt and pulled him from the plane, and put handcuffs on him.

'He was going for his pocket,' I told Dave.

He fished in Macquarrie's jacket and came up with a little brown bottle. It was heart medication. We weren't the only ones who were finding the situation stressful.

Even now, I have no idea what Pat Macquarrie intended to do once he reached Hamilton. He had been a pilot during the war, flying Spitfires and Hurricanes against the Japanese. He had a bit of history in the anti-apartheid protest movement. In 1978, he flour-bombed a softball game between a New Zealand team and a visiting South African team. But he didn't have anything aboard in the way of missiles or even leaflets on his way to Hamilton, so he wasn't going to drop anything. Perhaps he hoped just the threat of a deliberate air crash would be enough to force the game's abandonment, and perhaps he set down at Morrinsville because he had heard his mission had been accomplished—he was supposed to have had a transistor radio in the cockpit. I'm pretty sure he was lost when he decided to land. While I took part in a documentary about the episode, using my helicopter for a reconstruction, I never found out if anyone had learned what was going through his head. He was convicted of stealing the aircraft and a whole bunch of other stuff and slung in jail. Apparently, he was supposed to be flying the plane that flour-bombed Eden Park a couple of weeks later, but because he was locked up, Marx Jones took over instead. Macquarrie is dead now, so I suppose we'll never know.

***

There was far more to the Springbok tour than just rugby or even the morality of maintaining sporting links with a racist regime. While Muldoon and pro-tour supporters were forever arguing that politics shouldn't be mixed with sport, looking back, the 1981 tour was all about politics. Muldoon was hoping that taking a strong line on the tour would boost him in the polls, which had been looking a bit sick. He was right. But although he won the election later that year, the polls hadn't been lying. People were sick of Muldoon by now. The country was divided over all kinds of stuff, and not just rugby or even racism. People were sick of the regulation of every aspect of the economy and society. They were sick of paying high tax (those in the top tax bracket paid 66 cents in the dollar) and even higher interest rates (we were paying about 17 per cent on mortgages in those days). They were sick of unionism (or heavy-handed actions by employers). They were sick of Muldoon's generation, who had hung on to power for too long. Just about everyone in New Zealand, in short, was grumpy for one reason or another and plenty were ready for a change. The Springbok tour was really just an expression of all that.

Cometh the hour, cometh the man. A millionaire Wellington property developer by the name of Bob Jones decided that he would form a political party to contest the 1984 elections. Who knows whether he thought he would actually get elected: I doubt it. You can say what you like about Bob—or Sir Robert, as he is these days—but you can't say he's stupid. Probably he realised that any support he got was likely to come at the expense of Muldoon and the National Party and bring them down.

Regardless of what his ambitions were, his New Zealand Party did manage to attract a surprising amount of support, and although they didn't win a single seat in Parliament, they split the National Party's vote and Labour came to power. And when Labour began to implement practically every item on the New Zealand Party's economic liberalisation programme, Jones decided his party was redundant. Without consulting the rest of the hierarchy, he announced in July 1985 that the New Zealand Party would be going into recess. The decision was greeted with outrage by quite a few of his own supporters, but that didn't bother Bob: according to the best intelligence the press could gather, he had gone fishing on the Tongariro River.

And that's where I came in.

I got a call from Rod Vaughan from Television New Zealand who, along with a TV cameraman and a sound man, was keen to hire me and my chopper to take them looking for Bob. I assumed Bob was expecting them, so that's what we did. We did a few set-up shots for telly, where they filmed from the ground as the helicopter flew over and filmed from the air as we swept up the river. Then we went and did a fly-by of Bob's holiday home just outside Turangi, before heading to the river to look for Bob himself. There was an air of schoolboy excitement in the cabin, and that should have set alarm bells ringing.

We flew up the river until Vaughan said: 'That's his car, I reckon,' pointing to a car parked at the end of a track, whereupon I skimmed over the trees to the river. There was a lone angler standing in a rapid in the lower reaches of the river in a dark jacket with the hood up. We approached and hovered a little downstream from him. He reeled in his line keeping his back turned then stalked out of the river and into the bush. It occurred to me that he didn't look like someone who had been waiting patiently for a bunch of journalists to show up to interview him.

'Put us down,' Vaughan said.

So I set down on the open ground on the riverbank where Bob had entered the trees. The TV guys climbed out and began to walk in the direction he had taken. From the cockpit, I had a grandstand view of what happened next.

The cameraman Peter Mayo managed to swing his camera onto his shoulder in time to catch Bob lumbering from the trees. He made a beeline for the TV camera and briefly wrestled with Peter, who ended up in a heap on the ground. Then he rounded on Rod Vaughan and swung a pretty decent left hook right into his nose. He went after the sound man, too, but he wasn't sticking around.

'There's no need for that,' Mayo complained, as Vaughan stumbled away, his hands clapped to his face and blood oozing between his fingers.

'Now piss off!' Bob shouted, and with a last attempt to grab the camera and toss it in the river, he turned and walked calmly back into the bush.

Vaughan stood for a while letting the blood trickle down from the cut on the bridge of his nose and drip off his chin. Then he did a piece to camera. Everyone loaded up and we headed back to Turangi.

Of course it was huge news, and, in the papers just as in reality, Bob came out swinging. He had only defended himself against what he described as 'intolerable harassment'. The chopper's rotors had almost taken his head off, he reckoned. His only regret, he told *The New Zealand Herald,* was that he hadn't got the chance to punch the helicopter pilot, too.

Truth be told, the helicopter pilot was feeling a bit used. Jones's lawyer told the papers that Bob had been approached for an interview and had said, in no uncertain terms, that he was not interested. He had even attacked a *Sunday Star-Times* reporter and his photographer who had bailed him up at home that morning. I hadn't known any of this when I consented to fly Vaughan and Mayo in to find Bob on the river.

These days, TV and print journalists jump in helicopters every day of the week to get a story. Not back in 1985. The unholy alliance of the media and rotary wing aviation seemed to meet with general disapproval, not least from Civil Aviation, who came calling to see whether there was any truth in Jones's allegation that I had nearly chopped his head off with the rotors. I told them that they could see for themselves from the footage that I had been low, but not dangerously low, and that at no time was Jones or anyone else put in danger.

I was appalled a few days later to hear that I was going to be charged with dangerous flying. The matter was set down for the same day and in the same courtroom as poor old Foggy Fogden's. To have two pilots from the same company both up on charges of endangering life and limb with their helicopters wasn't a good look at all.

As if to make things worse, I was called upon to appear as a witness in Bob Jones's assault trial—for the defence! I had plenty of sympathy for him—I thought he had a pretty strong argument that he was being harassed by the journalists, and I felt no loyalty to them whatsoever—but I knew that anything I said might be used against me when I came to have my own day in court.

That's how it played out, too. When I was called, Bob's lawyer, a QC from Wellington by the name of Mike Camp, asked me whether it was understandable that his client had been frightened by my helicopter. I had no trouble conceding that given the Squirrel was one of the bigger machines around, and given helicopters are by their very nature loud and intimidating, it was perfectly possible to believe Bob had been scared by it hovering overhead. But when the prosecution tried to get me to say how low I was, I ducked for cover.

'How high off the ground were you, Mr Funnell?' he asked.

'Well,' I mused, 'It's difficult to say.'

'You must have some idea,' he said.

'How long is a piece of string?' I shrugged.

The prosecutor looked at me in frustration, but Justice Monaghan, the presiding judge, smirked slightly from the bench. Sensing a certain lack of judicial support for it, the prosecution gave up on the line of questioning.

Meanwhile, the Civil Aviation Division weren't letting go. We were summonsed to appear in court on 21 January 1986, where charges were going to be laid. When I was called, our barrister, Rob Vigor-Brown, apologised on my behalf for my absence, but cited the pressures on my time of my search and rescue work. We weren't actually doing a SAR job at the time, but the point was that I could have been called at any time. I was charged on a number of counts of the improper use of my helicopter. So was John Fogden, and we were both instructed to appear again in April for the trial itself.

The whole thing was a distraction. I don't think it actually affected our business—no one cancelled anything we had booked—but you never know who might have been deterred from chartering us because of the allegation hanging over us that we were a danger to ourselves and others. A bunch of emotional energy went into mentally preparing for the hearing, too. But just to prove the whole thing was a farce, I got a call from my solicitor, Mike Corry, about a week out from the day I was due to appear to say that the Crown were not going to contest the charges. The official reason given was that they had reviewed the television footage and concluded, with the assistance of an unnamed 'helicopter expert', that it would be hard to prove I was dangerously low. The other reason was that Bob Jones, who had been all set to be the star Crown witness, had declared that he was unwilling to say anything that would help the prosecution, given that my case was to be prosecuted by the same bloke who had successfully got a conviction against Bob. So I was off the hook. If you ask me, Rod Vaughan was pushing his luck chasing Bob down when he had made it clear he did not want to be interviewed. Bob dished out his own rough justice, and he paid for it with a conviction. Justice was done all round.

Poor old Foggy Fogden wasn't so lucky, and he was convicted on two charges of dangerous flying. I wasn't in court again—this time because I was on standby to pick up an injured climber from Mount Ngauruhoe—but Rob Vigor-Brown told me that Judge Monaghan had been openly scathing of the way the Crown had gone after Foggy, laying a raft of charges in the hope that one would stick. There was no clear definition of 'dangerous flying'. In the end, Foggy was fined $350. It was good to have it all behind us and to move on.

There was a sort of a sequel to the whole business six months later. In December, I got another summons involving Bob Jones, this time from the New Zealand Water Safety Council who were keen to use Bob's sky-high profile as an angler to make a television advertisement on keeping safe in rivers. They wanted me to hover above him so that he was caught in the downdraught, whereupon he would fall over. You would have had to have been hiding under a rock for the last 12 months not to have got the joke. I was happy to donate machine time and my expertise to my new best mate, Bob.

***

This wasn't the first time I had been charged with a flying offence. Back in the early days, I was called to attend an accident up on Turoa and asked to transport the patient to Palmerston North Hospital. I had returned to Turoa on my way home to drop off the skifield medical equipment, and with that done, I took off uphill. Some jumped-up Department of Conservation ranger decided I was showing off. He complained that I was flying below the legal permitted height.

Again the Civil Aviation Division was deaf to any of my explanations, so it went to court.

The Crown prosecutor told me that I could easily have taken off downhill so as to reach a greater height off the ground sooner, but I disagreed.

I explained that an engine failure would have been interesting if I was taking off downhill, because if I'd been trying to level the skids with the snow-covered mountainside, the helicopter would accelerate away. They're not called 'skids' for nothing. I also told him I had to factor in the recommended take-off profiles.

'You have to consider the Dead Man's Curve,' I said.

That got His Honour's attention.

'Dead Man's Curve?' he said. 'Forgive me, Mr Funnell, but you'll have to explain that one for me.' He asked me to approach the bench, which I did, producing a copy of the height–velocity graph I had brought along, which I proceeded to explain to him.

I told him that there was an accepted safe ratio of height to velocity. The recommended take-off profile was to build up airspeed before you sought height. You could safely go slowly if you were close to the ground, or you could fly slowly if you were high above the ground. When you plotted all this on a height to velocity chart, there was a shaded 'danger' area—the Dead Man's Curve. Operation in this area was to be avoided, as in the event of engine failure there would be no opportunity to perform a safe auto-rotation manoeuvre.

I'm not a particularly tall man, and in order to be able to show the judge my sketching, I had to stand on a little ledge at the foot of the bench. I was so engrossed in my explanation that when the judge signalled I could resume my seat, I forgot about the ledge. As I stepped backward, I lurched and crashed to the floor.

Mike Corry, my lawyer from Auckland, was on his feet in a flash.

'Your Honour, a practical demonstration of the dangers of a downhill take-off,' he said.

The case was dismissed.

***

The numerous other times I have appeared in court, they have all thankfully been as an expert witness. As I was fretting about the first of these occasions, someone gave me some excellent advice, to which I have always adhered since. Your job, they stated, is to be a completely independent witness. So long as you don't tell lies or stray outside the area of your expertise, you'll be fine.

A Rotorua solicitor with whom I'd had a bit to do called Brent Cooper once called me and asked if I'd appear as a defence witness for a client of his who had landed a plane untidily at Ohakea Air Base. I seem to recall his name was Jones, too. He had got hold of a de Havilland Devon, an old, twin-engined flying bus of a thing, and was taking a bunch of eight mates from Hamilton down to Wellington for the rugby sevens. He was a VFR (visual flight rules) pilot—qualified, that is, to fly only when he could maintain visual contact with the ground—and the weather forecast wasn't flash. He probably shouldn't have flown, but the pressure from his mates was probably too much for him. He decided to go.

The weather closed in from the west as they neared the vicinity of Hunterville, and the cloud base lowered. With the benefit of hindsight, he might have been okay if he'd turned back then, but then again, the weather was likely closing in behind him, too. In any case, he chose to press on. He realised he was in trouble shortly afterward, and called Ohakea for some radar assistance. There's a nice, big, long runway at Ohakea, so it was a good choice to make a landing with lots of safety margin. Jones was probably keen to get out of the air by now. He made a downwind approach—it's safer and more prudent to land into the wind—and as the aircraft landed, the undercarriage collapsed and the plane sledged to a halt amid a shower of sparks. No one was injured, and the boys—Jones excluded—carried on to Wellington by bus.

Civil Aviation wanted to throw the book at the young pilot. The defence was arguing that there was some fault with the undercarriage that caused it to collapse, but I told Brent that if directly asked what I thought had happened, I would have to say that I thought the difficulty in gauging airspeed in a downwind landing had caused Jones to stall the plane and drop it on the tarmac.

Because I was a witness for the defence, my involvement in the case came very late in the piece. Brent sent through transcripts of each day's action which I read in bed to keep abreast of developments. Finally, my day in the High Court in Hamilton came.

Brent is one of those lawyers who loves to perform in front of a jury. He's a charmer, and he really turns it on for them. When I was put in the witness box, he asked me my name, about when I had got my pilot licences, about how many hours I had accumulated, about how many of those hours had been earned while engaged in search and rescue work...

It was getting bloody embarrassing, and I was squirming. Judge Wolff, who was presiding, noticed and intervened.

'Mr Cooper,' he said, 'I am sure that, like me, everyone in this room has heard of Mr Funnell and his exploits. I think we are all satisfied of his credentials to appear as an expert aviation witness in this case. So let's just move on, shall we?'

You could tell Brent didn't like missing his chance to paint the full picture for the jury, but he had no choice. He led me through the weather and the options that the pilot would have had. Then it was the prosecution's turn.

I had introduced myself to the Crown prosecutor, Rebecca Mann, at the beginning of the day's play, which seemed to have taken her aback. She started out nice.

'Do you know the defendant in this case, Mr Funnell?'

'No,' I said. 'But I'm guessing that's him there in the dock. The one who's not dressed in a green uniform.'

Laughter rippled through the court, and the jury looked grateful for this moment of light in an otherwise pretty dull day. Brent Cooper beamed at me. Ms Mann looked decidedly unimpressed.

'Mr Funnell, I have a map of the flight path here, as recorded by radar on the day in question. I wonder if you'd mind having a look at it.'

She brought it to me where I was standing in the witness box, and I leaned forward to get a better look. I had been scrupulously careful in turning off my cell phone and my pager before I entered the court, but I must have leaned on my pager, because soon after Ms Mann walked away and I sat down, it suddenly pealed into life.

Brent was on his feet in a flash.

'Your Honour, please excuse Mr Funnell. He is an air ambulance pilot who has been responsible for a large number of dramatic and successful rescues in his distinguished career. This could be a very important call, a matter of life and...'

I was trying to add my apologies, too, telling Judge Wolff what must have happened. He looked unconcerned.

'I understand, Mr Funnell. Please sit down, Mr Cooper.'

Brent sat down, but looked quietly pleased with himself that he'd managed to get a few more brushstrokes in.

Ms Mann proceeded with her cross-examination, and was doing her damnedest to get me to say that Jones had acted illegally or recklessly. I took care to tread a line between agreeing with her and advocating for the pilot. It was nerve-wracking stuff, and I was feeling quite beaten up when 11am rolled around and it was time for the morning adjournment.

His Honour explained that the court was going to rise for morning tea, and said that if counsel for the prosecution and for the defence had no objection, he would like to see Mr Funnell in his chambers.

Here we go. I'm in the shit now, I thought.

I sent a mute mayday to Brent. He smiled, and said: 'No objection whatsoever.'

Ms Mann didn't have a problem, either, so as the judge left via the back door in the richly wood-panelled courtroom and everyone else filed out, I sat there.

Eventually, the Clerk of the Court, who had stayed in his position, said: 'Come on, Mr Funnell. His Honour is waiting.'

He led me through the door. The judge was waiting just on the other side, at the near end of a corridor. I had no idea what happened on this side of the back wall of the court.

'Come with me, Mr Funnell,' he said.

'Your Honour, I feel like an errant schoolboy following the headmaster to his office,' I said.

'There aren't any canes in my office, Mr Funnell. Come on!'

He showed me into a pleasantly furnished office at the end of the corridor and sat me down. The Clerk of the Court joined us.

'Now, Mr Funnell,' Judge Wolff said. 'The Clerk has joined us to ensure we don't talk about the case. I expect you're wondering why we're here.'

'If it's about my pager going off in court, sir...' I began, but he waved dismissively.

'I called you in, Mr Funnell, because I want to thank you,' he said.

That shut me up. He explained that Mike Bannerman, the pilot of an aircraft that had crashed in Christchurch in 2003, killing seven scientists from Crop and Food Ltd, had a brother who had been at law school with him, and was a personal friend of his. I had given the Bannerman family quite a lot of support in the aftermath when everyone rushed to blame the accident on pilot error. Mike's brother and wider family were immensely grateful, he said, and so was he.

I was incredibly relieved. We had a convivial cuppa together, where we talked about mutual acquaintances—His Honour had grown up near Hunterville—and then, after a while, he glanced at his watch.

'Christ! Is that the time? We'd better get back in there, Mr Funnell.'

The case wound up quite soon afterward. The jury found Jones not guilty, whereupon Brent Cooper applied for costs. When Judge Wolff handed down his judgment, he was scathing towards Brent. In fact, he had a pretty low opinion of everyone in the case, it seemed, except me. There was a line in there that read something like: 'The only credible witness in this whole case was a certain Mr Funnell.'

Brent shrugged off the roasting he'd got. He was over the moon with the result.

The Philips Squirrel rescue helicopter.

Attending another motor vehicle accident.

CHAPTER SEVEN

CAP IN HAND

We were doing hospital transfer work more and more frequently, and one day while I was talking to Dr Rowan Hyde, the discussion strayed to the possibility of setting up a rescue helicopter based in Hamilton. I knew where we could get a suitable machine. Grant Biel, my friend in Auckland, and his wife Billie owned a couple of helicopters: one they used pretty much year round; the other, an Aerospatiale Squirrel like the machine we were using for Philips Search and Rescue Trust work, they tended to lease out each summer for work in Antarctica. I thought we'd be able to work out some sort of deal for at least some of the year to get sponsors supporting the operation. We also had a pilot. Grant Bremner was a Taupo boy who had gone offshore for a while, flying mostly in Papua New Guinea, but he hadn't liked it and had returned home looking for work. I had him flying part-time for me, doing odd jobs while he waited for something more permanent to come along. He would be just right to fly the Hamilton-based machine.

But helicopters don't come cheap. Every time you pull the collective, money starts draining from your account, what with the costs of fuel and the hours you burn off your machine with its rigorous maintenance schedule: for every 100 hours' flying time, you have to have a full inspection done. Every 600 hours, an even more painstaking inspection is carried out. Even if you're not working your machine very hard, it gets minutely inspected every six and 12 months, and once you've done 3500 hours on an engine, it has to be fully rebuilt. None of that is cheap. Looking at the books of the Philips Search and Rescue Trust, I reckoned we would need no less than $100,000 per year to run another chopper. Dr Hyde and I talked to a number of well-meaning groups who were keen to have a locally based, dedicated air ambulance helicopter in the Waikato to see if we could attract any sponsorship. We did lots of talking, but a few months along we had nothing to show for it.

'This stuff is quite hard,' I admitted to Dr Hyde.

'There must be people who do it professionally,' he said. 'Let's try and find one of them.'

Asking around, I was given the name of Hugo Percy, whose company Blue Sky Promotions seemed to specialise in public campaigns of one sort or another. Hugo was a charismatic sort of character, full of energy and enthusiasm, and he was very confident he'd be able to raise the sort of money we needed. So, buoyed by his assurances, Dr Hyde and I began to recruit support amongst the other hospital doctors for the concept of a Waikato rescue helicopter. Most were in favour: Dr Tony Cull, head of the neonatal unit, and Dr Jack Havill, head of the intensive care unit, were particularly keen on the idea. Doctors Cull and Rowan, and I approached Jack Jenkins, the chair of the Waikato Hospital Board.

Unlike many of the people who occupy these posts, Jack was a dynamic individual, and when he decided something should happen, he made sure it did happen. So it was with the Waikato helicopter. I think it only took one meeting—'Great idea!' said Jack—and I was on the phone to Grant Bremner explaining the arrangements.

'We'll be using Grant Biel's Squirrel,' I told him, 'and we'll be using the tennis court next to the nurses' home as a pad. Jack Jenkins has told us we can just get rid of the net and take it over.'

'Righto,' Grant said.

'And you'll be living in the nurses' quarters,' I added.

'With the nurses?' he asked.

'With the nurses,' I confirmed.

There was a silence on the other end of the phone, where the young helicopter pilot processed this information. I had the strong impression he was thinking that whatever disappointments he had suffered in his flying career, he was confident this latest move would not be another of them.

***

The Biels were big supporters of what we were doing with the helicopter at Taupo, and they had no hesitation in offering their machine to the Waikato branch of the operation for three months, completely free of charge. So now we had a pilot, a machine and somewhere to park it. We didn't have anywhere to store the gear we needed, but when we approached the local Toyota franchise holder, Terry Booth, he was delighted to come on board. He supplied a brand-new Toyota Hiace van, which we got signwritten with the words 'Terry Booth Toyota' and 'Air Ambulance'. That took care of the gear storage, and it also gave Grant something to get around in. But even though we were now in business and actually flying missions, it was all by the seat of our pants. We had only basic gear, both for the chopper and for the rescue and transfer work: this severely limited when we could fly (such was the state of technology back then that we were largely restricted to daylight hours), and what we could do when we got where we were going. We badly needed more money, but every time I asked Hugo how he was going in tracking down sponsorship, he would express the same unwavering conviction that he was going to land something 'next week'. A few weeks of this and I was beginning to get disillusioned. We were carrying the cost of the fuel for the chopper and for the van ourselves, and we were paying Grant's wages.

We weren't the only ones who were disappointed.

'How's it working out with the nurses?' I asked Grant. 'Are they missing their tennis court?'

'Nah,' he said. 'They never use it. But they're a pretty grumpy bunch. If you take off when they're trying to sleep, they let you know all about it when you get back.'

'Still,' I said. 'There must be compensations?'

He shook his head sadly.

'Bloody nurses don't do anything but work and sleep,' he replied.

***

We had been operating for about six months when Grant Biel phoned and apologetically told me that he had been offered a contract to send the helicopter to Antarctica again for the summer. He was happy to let me continue to use it, but he didn't think he could let me have it for free any longer. After all, our original agreement had been for three months, just while we got our funding in place. He offered to lease it to us at very favourable rates, but, of course, without some source of funds, it wasn't going to work for us at all.

I rang Dr Hyde and passed on the news, along with my frustration with Hugo's failure to deliver.

'How hard can this stuff be?' Dr Hyde said. 'Let's go and see the big boys in Hamilton business ourselves.'

We used the list of organisations Hugo had targeted, and, naturally, given we were looking for a locally based entity with deep pockets, we ended up with Trust Bank Waikato at the top. I found out that the general manager was a man by the name of McRobie. I phoned and made an appointment.

So a few days later, the doctor and the helicopter pilot brushed the lint off their best suits, gave their shoes a polish and drove to the Victoria Avenue offices of Trust Bank Waikato. We took the lift to the third floor, and the doors opened on an impressive space, a wood-panelled foyer with a lofty stud, a pair of big wooden doors (firmly closed) and a large desk behind which a lovely young lady was seated. She gave us a cool smile as we introduced ourselves and asked us to take a seat. She picked up a phone, and simply said: 'They're here.'

We sat there awkwardly, both reflecting that neither of us had ever done anything like this before.

After what seemed like ten minutes, the double doors opened and we were ushered into the presence of the general manager of Trust

Bank Waikato, who turned out to be a smallish, immaculately dressed man with greying hair, seated behind an even bigger desk and with an even bigger set of double doors (firmly shut) behind him.

We introduced ourselves to Mr McRobie, and he gestured to the seats in front of his desk. The rich leather of his own seat squeaked as he settled himself into it.

'Gentlemen,' he said. 'How may I help you?'

Dr Hyde and I exchanged nervous glances then I began to speak.

Mr McRobie listened politely, nodding occasionally, as I told him about the rescue helicopter, the lease offer, the operating costs and the impossibility of keeping it running without a reliable source of funding. Dr Hyde spoke about the importance of the helicopter to the medical outcomes of patients transferred from outlying hospitals to Waikato.

Mr McRobie asked a few questions that showed he was both interested in and engaged with what we had been saying. Every now and then, he would write something down with a heavy-looking gold pen.

'How much did you think you would need?'

Dr Hyde and I exchanged glances again.

'We think we could get by on $150,000,' I said. 'Per annum.'

Mr McRobie's face betrayed nothing. He set the pen down and stood up. We stood up, too. He walked around the desk, opened the doors and stood there with the clear expectation that we should leave.

'Thank you, gentlemen,' he said. We thanked him back and shook his hand. The doors closed behind us. The receptionist showed us back to the lift.

'Mr McRobie will be in touch,' she said.

We were pretty despondent in the lift. This stuff was hard! Our respect for Hugo Percy had just ratcheted up a few notches.

'Who's next on the list?' Dr Hyde said.

We parted, agreeing that we would need to devise a more effective strategy before we approached our next potential benefactor. But when I got home, I phoned Grant Biel and told him that it looked like he would soon have his chopper back.

'Hang on to it for now,' Grant said. 'Keep trying with the funding.'

I was incredibly grateful. Who knows what pressure it was putting on Grant and Billie and their family to let us have the machine gratis.

***

Dr Hyde and I had been polishing our pitch and were just about ready to try our next-in-line option when I got a call from Mr McRobie's secretary.

'Mr McRobie would like to see the two of you again to discuss a few points. Would you like to make an appointment to come in?'

I made the appointment for a few days' time, and phoned Dr Hyde. This, we agreed, looked vaguely promising.

So the doctor and the helicopter pilot togged up again, drove to Victoria Avenue and rode the elevator to the third floor all over again. This time, the lift doors opened on a much warmer smile from the receptionist.

'Take a seat, gentlemen,' she said. 'Would you like a cup of tea?'

This seemed like an unequivocally good sign, but we were still too nervous to accept the offer of tea.

After a few minutes, the big doors opened again, and this time Mr McRobie himself was standing there. Beyond him, the other set of doors had been opened onto a cavernous, beautifully appointed room with a highly polished table in the centre that seemed to stretch to vanishing point. It was flanked with luxurious-looking leather chairs, two of which were occupied by men in expensive suits. We were introduced to Mr Grey and Mr Galbraith, both Hamilton solicitors and members of the board of directors of Trust Bank Waikato.

'Gentlemen, Mr Grey and Mr Galbraith would like to hear your story,' Mr McRobie said.

Dr Hyde and I delivered our pitch. It was a bit better than last time around, as we'd put a lot more thought into it. We told a few stories that our distinguished listeners may have noticed in their newspapers—and we were cunning enough to have chosen stories featuring kids, on the understanding that stories about children had more emotional impact. We told them that the local branch of St John's was interested in what we were doing: Gary Salmond, the Waikato head ambulance officer, and his deputy Eddie Jackson were supporters. The Accident Compensation Corporation was fully behind us, recognising the importance of early intervention in medical emergencies, and of the timely delivery of patients—particularly the victims of head and spinal injuries—to treatment.

The three men listened attentively, giving nothing away. We finished and sat there like a couple of startled rabbits. We were country boys, after all, and here we were out of our element in the citadel of wealth and power. I was suddenly convinced that our plea for the bank's money must have sounded ludicrous to these sophisticated men.

Sure enough, a sombre expression settled on Mr McRobie's face.

'Gentlemen, I know you asked us for $150,000 to operate the helicopter for a year. I have some news. We're not going to give you the sum you have asked for.'

His words were like a physical blow. It took me a moment to realise that he was still speaking.

'Instead, we're going to give you $120,000 per annum for five years. I know that falls short of what you need, but we are also going to give you a list of names of businessmen in Hamilton and you will approach these men to make up the shortfall. I expect you will find them receptive to your approach. In recognition of our support, we want to see the bank's name prominently displayed on the helicopter.'

I mumbled that I was sure this could be arranged, still not quite believing what I was hearing.

Everyone stood up, and hands were shaken all round. The beaming secretary showed us to the lift doors. We waited quietly as the lift arrived and the doors opened. We stepped inside and kept our expressions neutral as the doors closed, shutting off our view of the room and its occupants, and the lift lurched into motion. I recall we managed to wait nearly until we reached the ground floor before we turned to each other and yelled: 'Yeehaa!'

'Un-bloody-believable, doctor!' I said.

When the lift opened and we emerged, flushed and triumphant, we were greeted with surprised looks. I'm quite sure they heard us three floors up in the hush of the boardroom, too.

***

It was the start, as they say, of a great relationship. It may have begun with frosty formality, but I became firm friends with Ross

McRobie, and over the years he was a tremendous supporter of the helicopter. He even took part in our promotional activities. There was one particular television advertisement where he stood in a field and said to camera: 'If you want an overdraft, come and see me.' His hair and clothes start flapping wildly and the camera shot opens out to show the helicopter, emblazoned with the Trust Bank name and logo, hovering above him. Overdraft, get it?

The other funders towards whom Ross steered us were just as forthcoming as he indicated they would be—John Gallagher of Gallagher Group, the Perry Foundation and others all came aboard, and Terry Booth kept giving us a new Toyota van each year, but the bank remained our principal funder.

Like all close relationships, it had its ups and downs, of course. Costs were forever rising, and each year we had to go back, cap in hand, to try to finesse another $10,000 to $20,000 out of the bank. Year in, year out, they would oblige, to the point where we began to feel as though the annual increase was ours as of right. Then, one year, we got a bit of a shock. We were back running on the bones of our bums by now—Grant Bremner was finally taking a well-earned day off and I was filling in. Little old ladies were holding cake stalls and flogging knitting at fairs to try to scratch up the cash we needed. Everything was compromised to a greater or lesser extent. And into this fraught, cash-strapped climate, the bank dropped a bombshell.

The first inkling we had of it was when the double doors on the third floor opened and it wasn't Ross McRobie, but a man who introduced himself as Bruce Cronin, the bank's deputy general manager, who was there to greet us. He told us that there would be no increase that year.

'What do you mean?' I spluttered. 'Where's Mr McRobie?'

'Mr McRobie and the directors have decided,' Cronin said haughtily, 'that we get better value out of our sponsorship of Waikato Rugby than we do the rescue helicopter. So he regrets to inform you that the bank has decided to direct more of its funding to that area.'

We were thunderstruck, and we set David Wickham, the trust secretary, to writing letters to ask Mr McRobie to reconsider. Eventually, Bruce Cronin wrote to say that Mr McRobie had heard us, and to demonstrate the wisdom of its decision, the bank had commissioned a public relations firm to survey the people of the Waikato region. We were cautiously optimistic about the outcome, but nervous nonetheless.

The next call we had was brusque.

'There's no need to come in,' Bruce Cronin said. 'Your application for an increase in the level of your funding has been approved.'

'Thank you,' I said. 'Any chance we can see the results of the survey? You know, just out of curiosity?'

'That won't be possible,' Cronin replied. 'The results are confidential.'

Years later, upon the occasion of Ross McRobie's retirement, I told him that I had always been curious about the results of the survey.

'Funnell,' Ross sighed. 'I knew you of old. I knew from bitter experience that if I gave you and the doctor an inch, you'd take a bloody mile. We commissioned that survey expecting to find that Waikato Rugby was giving us excellent value for our sponsorship dollar. Well, the survey showed that every man and his dog knew about the Trust Bank rescue helicopter. Only some people knew about the rugby. But there was no way in hell we were going to give you and the doctor that information!'

***

Funding wasn't our only headache in those days. We also had our share of attention from officials high and low. By the time the Philips Search and Rescue Trust had been operating for about 18 months, we were getting around six call-outs a week, sometimes more. On 23 June 1986, a man was transferred by helicopter from Napier Hospital to Wellington. He had a stab wound that had been healing well before he decided to play soccer. He collapsed, and was in a bad way in hospital when the helicopter arrived. There was no helipad at Napier Hospital. There were a couple further afield, but it was judged too risky to move the patient to either of these, so the machine set down right outside the front doors of the Emergency Department. Naturally enough, this made quite an impression on the various bystanders, as it did about a month later when we arrived, with four-year-old Kirsty McVicar aboard. Kirsty had been with a camp group at Mount Ruapehu and was tobogganing on a plastic bag when she sailed into a depression in the snow and had a hard landing. She was made as comfortable as she could be in the skifield medical centre. When we arrived, Pat Wynne pronounced her lower leg very badly broken with the circulation possibly compromised. To get the proper treatment and stand the best chance of saving her leg, she needed to get to one of the larger hospitals quickly. We chose Napier, as that is close to where the McVicar family live. Poor Kirsty had already suffered plenty, and she wasn't keen at all on being loaded in the helicopter. Once she was installed, she looked around her, terrified, and she was on the verge of panic when we lifted off.

'Relax,' Pat told her. 'It's all quite safe. There's nothing to worry about at all. I've got a secret signal worked out with the pilot. If there's any trouble, he touches his head.'

Kirsty smiled bravely. Twenty minutes later, she was just about calm—and that's when I felt a sudden need to scratch my head!

We set down by the front doors. I remember Kirsty clearly, the way I remember all of the kids. She was a brave little girl. I've met her in more recent years, and she still walks with a limp. But without Pat's expert care and the capability of the helicopter, she might not have kept that leg. Her father Garth, who is quite well known in his role as spokesman for the Sensible Sentencing Trust, is a friend.

As luck would have it, the cold spell that hit the Central Plateau the weekend Kirsty injured herself had us back at Napier again two days later. Kirsty's accident was on the Saturday; on the Monday morning, a car hit a patch of black ice near the Rangitaiki tea rooms, 30-odd kilometres east of Taupo, and slid off the road. There were no injuries. Shortly afterwards, a van hit the same patch of ice and crunched into a tree. Again, any injuries were minor. Almost immediately, a third car hit the ice and sailed over a bank. The occupants were bruised and shaken, but otherwise unhurt. As everyone was standing around congratulating themselves on their lucky escape, yet another car arrived, hit the ice, slid into the van and then slammed into a bank. The young driver was relatively unhurt, but his mum and grandmother, who were both in the car, were seriously injured. When Pat and I arrived, he decided the mother would be okay travelling by road to Taupo, but May Waters, the elderly back-seat passenger, had sustained severe head injuries and had to go to Napier.

This time, someone must have got word we were coming, because there was a photographer from the *Hawkes Bay Tribune* there to take photographs as I lowered the Squirrel into the tight space between the car park and the main doors of Napier Hospital. So our picture was in the paper the following day, and a few days later our 'recklessness' was being condemned in a meeting of the Napier City Council by councillor Alan Styles. There had been an epidemic of such landings at Napier, he reckoned, and we were only doing it because it was 'fashionable' for show-offs like me to land right outside hospitals. The mayor, Dave Preverton, agreed: we had breached any number of Civil Aviation rules and regs, and he meant to lay a complaint. No one troubled to ask us why we made the decision to land where we did on each occasion: never let the facts get in the way of a good, old-fashioned expression of outrage! We responded in the papers, and the families of May Waters and Kirsty McVicar weighed in with their gratitude, too. The fuss died down.

I suppose we were a bit cavalier with the rules and regulations at times. We certainly didn't always have the luxury of obeying the letter of the law when it came to when and how you flew in poor light or bad weather. We tended to under-report (on the principle that if it wasn't written down, it never officially happened) the instances when we flew in bad conditions, or failed to maintain the statutory distance from cloud. In fact, we generally took a 'what they don't know can't hurt them' attitude to the Civil Aviation Authority; it was only really when the newspapers carried reports of what we had done that we were diligent in sending the paperwork to Wellington, because CAA had a clipping service. Even so, we were as economical with the truth of the matters as we felt we'd get away with. I'm sure CAA weren't fooled in the slightest.

Civil Aviation has a job to do. After all, they're there to ensure that in trying to use helicopters to help in the case of emergencies we don't become an emergency ourselves. But some of the personnel there were more officious than others. One who sticks in my mind is a man named Stuart Smith. Stuart had an engineering background. I knew him quite well, but he began to get up my nose when he seemed to take a particular interest in what we were doing and how many rules and regs we were violating to do it. He wanted written notification containing full details of every single infringement.

'Bloody hell, Stuart,' I said to him on more than one occasion. 'All we're trying to do is save lives.'

'Ah, yes,' he'd reply. 'But you can't go breaking the rules to do it.'

He was only doing his job, and his job was to test our decision-making processes. But he was testing them too rigorously for our liking. It seemed the only way we'd get Stuart off our backs would be to show him the way things worked. I invited him and his family up to Taupo for a weekend.

'You can have a first-hand look at our operation,' I told him. 'There's bound to be a job. You can come out and observe.'

Grant Bremner was aware of the difficulties I was having with Stuart. So when Stuart came up to stay, Grant was under instructions to let me know if anything was happening anywhere in the region—Tokoroa, Whakatane, didn't matter: we'd drive to catch up with the action.

As it happened, on the very first day that Stuart was with us, a Friday afternoon, Grant called and said that he had the Waikato intensive care team aboard and was en route to Taupo. Grant had asked the medics if they minded me and a CAA observer attending, and they were happy.

'Here we go,' I said to Stuart. 'We're off.'

'Where?' he asked.

'The hospital. They've got a case that they urgently need to transfer to Waikato.'

We drove to Taupo Hospital and arrived just as Grant was putting the machine down on the helipad. The doctor and the nurse comprising the intensive care team climbed out and began unloading equipment. Those bags and the monitoring equipment that the ICU team travels with were our ticket into the A&E ward at Taupo Hospital, so Grant and I each grabbed something. I gave a bag to Stuart. We trooped after the medics into the hospital and into the emergency room.

Stuart was pretty pale of complexion at the best of times. As the doors swung open on the tight knot of bloodied people clustered around a gory gurney and standing on a blood-stained floor, he went as white as a blank sheet of paper.

Grant and I began setting up equipment.

'What's the story?' I asked.

'Suicide attempt,' a doctor replied. 'Young fulla here put a .22 rifle in his mouth. He's alive, but he's messed himself up a bit.'

That was an understatement! He had blown off most of the frontal lobe of his head. The doctors were doing their best under difficult conditions to clear the airway so that they could intubate him.

One of the nurses gave me a hard look and glared pointedly over my shoulder. I glanced at Stuart, who had backed off into the corner of the small room and was huddled there miserably.

'What's wrong?' I asked him.

He opened his mouth to speak, but closed it quickly and swallowed hard.

'Oh, Christ,' I said. 'I worked pretty hard to get permission for you to be in here, Smith. Don't you go collapsing on me. That wouldn't be a good look at all.'

Stuart kept it together, but he was very quiet as the patient was finally intubated, attached to various drips and monitors and loaded on the stretcher ready for the ride in the chopper. He looked highly relieved as we wheeled the guy out. But just outside the building, the balloon that sealed the tube in his airway collapsed, and there was a sudden flurry of activity. The stretcher was turned around and rushed back to the emergency room so that a new tube could be inserted down the patient's throat.

This was too much for Stuart. He waited outside while all this was going on.

It was early evening by the time the chopper lifted off and beat its way north toward Waikato Hospital. Stuart and I climbed back in the car. Stuart was still pale and breathing hard. He was very quiet on the short ride home, but a bit of colour had come back into his cheeks by the time we arrived.

'Good timing,' Trish said, as we walked in the door. 'Dinner's ready. It's steak!'

'Oh, Jesus,' muttered Stuart, and sat with his face averted from his plate.

'Not eating?' his wife asked.

'Nah,' he gulped. 'Not really hungry.'

I helped myself to half of his steak.

***

The following morning, I had jacked up with Rowan Hyde for Stuart and me to visit and talk about the work of the rescue helicopter. Stuart had rallied overnight and was up for the trip.

Dr Hyde met us at the doors to the ICU. I was well accustomed to the sights that greet you inside these places, but it was plainly all new to Stuart. There were nine beds, all occupied. Each patient was variously intubated and wired up to drips, monitors, pumps, suction units, oxygen, and each had his or her own nurse. They call it 'intensive care', and it's intense, all right.

Rowan was one of two doctors on duty. He shook hands with Stuart.

'Come and have a look around,' he said.

He gestured at the first bed.

'This patient came to us by helicopter from Whakatane. Motor vehicle accident. Serious head injuries. Pretty lucky to be alive, really...'

'What happened to that patient from Taupo last night?' Stuart asked.

'Oh, we'll get to him,' Rowan replied. 'This patient came to us from Taumarunui. Another motor vehicle accident. Chest injuries this time.

Wouldn't have made it to hospital in the old days.'

'Is that the Taupo guy over there?' Stuart pointed.

But Rowan wasn't to be hurried.

'Getting there. Getting there,' he said. 'We're a bit worried about this poor fulla. He's another road accident, and he's got pretty major head trauma. Wouldn't have stood a chance without the helicopter.'

And so it went. Finally, we stood at the foot of a bed in which a slight frame lay, its head swathed in bandages.

'This is your guy. Twenty-one years old. Gunshot wound, major head trauma.'

Stuart studied the face. It was clean and pink, but the eyelids were swollen and bruised.

'It doesn't look a thing like him,' he said.

'Oh, we cleaned him up and we'd describe his condition as stable right now,' Rowan said.

'What's going to happen to him?'

'Well, he's performed a pretty crude frontal lobotomy on himself, so he'll wake up with a headache and a self-inflicted change of attitude. We'll manage the headache for him. He should be okay.'

'What would have happened if the helicopter hadn't been available?'

'Oh, he would have died, for sure. They do a pretty good job at Taupo Hospital, but they don't have the resources there that we have. This is a major insult to the brain, and he needs intensive, one-on-one management. They just don't have the staff or the resources away from here. They might have managed in Rotorua, but there would have been a one-and-a-half-hour ride in an ambulance ... touch and go, probably.'

You could tell our message was beginning to strike home with Stuart.

'So, what if the helicopter couldn't get through because of weather or something?'

'Dead,' said Rowan flatly. 'He would have died on the table down in Taupo, or in an ambulance en route to here or Rotorua.'

'So you're pretty reliant on the helicopter.'

'It's absolutely critical to saving the lives of people like the people in this room.'

As he and his family headed back to Wellington, Stuart shook my hand.

'Thanks, Funnell,' he said. 'I've got a much better understanding of the importance of the work you're doing.'

Of course, that wasn't quite the end of our troubles with CAA. But they noticeably backed off, and whenever we did receive a 'please explain' letter, it generally wasn't Stuart's signature at the foot of it.

Near white-out conditions on Mount Ruapehu.

CHAPTER EIGHT

TO THE LIMITS

With the funding of the Philips Search and Rescue Trust behind us—and we were constantly raising funds through radio appeals and raffles and so on—we were becoming much more professional in our search and rescue activities. I knew from experience that one of the essential requirements of that kind of work, to say nothing of commercial helicopter activities in general, was communications, so I had employed Dean Sparks, who was running his own radio shop in Taupo, to take care of the comms side of things for us. He set us up with a very effective radio network, using handhelds as well as base sets in cupboards in the hangar and at home, along with a repeater station we'd plonked on a hill at the far end of Lake Taupo. One day, we were sitting in the smoko room when an officious-looking chap entered and started asking questions about radios. He was an inspector, he told us, and he believed we were operating unlicensed radio equipment. Well, of course, we were. The requirement to license every single transmitter was plainly ridiculous, so we hadn't bothered. We were busy denying it when the radio in the cupboard burst into life.

The inspector looked at us in triumph.

'Open the cupboard, please,' he ordered.

'No,' I said, and folded my arms. There were 10 of us and one of him. He left, all but waving his fist. And a few days later, when the hangar was unattended, he came by and emptied the cupboard of radios. We did our best to shame him in the press—we used the radios for saving lives—but he wouldn't budge. So rather than pay for a whole bunch of licences, I got Dean to build us a new clandestine network. Those were the battles you had to fight in those days.

***

We'd been joined in 1986 by a young policeman by the name of Barry Shepherd, a lean, wiry sort of bloke with icy blue eyes and a naturally sceptical cast to his features that must have been a real asset in his work as a policeman. But search and rescue was his passion, and he began to come out with us more and more. He really proved his worth on a harrowing job we did in the Kaweka Ranges in July 1990, where a Cessna 172 with four young people aboard crashed en route from Napier to the Central Plateau on a sightseeing expedition. The plane was reported overdue around 12.30pm and a locator beacon signal was noted by an Australian station at 1pm. A

Hawke's Bay-based helicopter found the wreckage teetering on the brink of a steep gorge in dense bush. They were able to get a crewman onto the ground in order to establish that two were dead, two were alive, but they weren't equipped to extract the victims. Pat, Barry and I flew there, and with Barry on the winch, we put Pat down. He had to cut a 30-metre track through the bush to get to the plane, which was largely intact. The two deceased—a 19-year-old woman and a male passenger of similar age—had been thrown clear, but the 23-year-old pilot, Alex Williamson, and a female passenger, 19-year-old Michelle Batt, were still in the plane and alive. Pat stabilised both as best he could, and chose to winch Michelle out first. We were using a new stretcher system—a light, flexible thing that you sort of wrapped around the patient—and it wasn't a success. The downwash funnelled through the four corners and it spun like a propeller. The winch cable was twisted like mad by the time we got her in the cabin.

Pat stayed with Alex as we flew Michelle to Hastings. We got there at 5.30 as the light was failing. With Michelle Batt in the hands of the medics, we dashed back to the ranges where we performed a very difficult extraction in the near-dark. We got Alex Williamson to Hastings by 6.15, but it was too late for him. He died soon after reaching hospital. We were exhausted by the end of it all, but proud of the way we had worked together. And over many years after that, Pat, Barry and I formed a formidable team.

Practice, as they say, makes perfect, and as time went by and we attended more and more incidents, we got better as a team at what we did, too. In its first year of operation, the Philips Search and Rescue Trust was called upon to perform 32 search and rescue operations. The nature of the place in which we lived meant there would never be a shortage of that sort of work: the Central Plateau is the North Island's playground, really, with the volcanic massif and the forests around Taupo luring hunters, trampers and skiers, and the lake and rivers drawing fishermen and boaties. Forestry was big business in the area, and even with today's high health and safety standards, it's still a risky occupation. Back then, when men didn't wear protective clothing and gear, it was a miracle we weren't called upon more often to pick up the pieces of forestry workers. And then there were the roads, which are capable of catching unwary and inattentive drivers out even when they aren't wet or frozen, which for half the year they are.

Most of the time, we were able to record positive outcomes for the victims whom we were trying to help. But not always, and some of those jobs have a way of sticking in your mind.

Late one summer evening in 1986, as I was sitting listening to the rain pelting the hangar roof, I got a phone call. It was the police. They had learned of an abseiler in difficulties in the Bay of Plenty, and they wanted a chopper with a winch and a cliff-face rescue team to attend.

We quickly set off with Pat Wynne, touched at Turangi to pick up a specialist cliff rescue team, and headed for the Bay of Plenty. It was pretty obvious as soon as we were off the ground that we would be lucky to get over the Kaimai Ranges to the scene. We turned left and tried to find a way over where the road follows the Wairoa River through the hills. No go. We turned right and had a look in the vicinity of Rotorua, but that was no use, either. So the decision was made to set down at Tokoroa so that the cliff rescue boys could drive on to the scene while Pat and I could get a bed at the hospital and wait for the forecast clearance in the morning.

I learned a few more of the details. A couple of blokes from the Bay, Mike Briggs and Mike Avard, had headed into the area south-west of Te Puke to abseil down a feature known to local climbers as Gibraltar Rock. Both were experienced, and conditions had been fine when they set off. But as the afternoon wore on, it clouded over and started to rain. At some point as Mike Briggs was sidling across the face above Mike Avard, he dislodged a rock the size of a table-top. He didn't know whether it was this slab or other debris that it loosened as it fell, but Avard had been struck in the chest and legs and badly hurt. The worst of his injuries was a shattered hip: his femur had been forced through the joint and was protruding from the skin. Briggs had managed to secure him to a narrow ledge and made him as comfortable as he could, and had then made his own slow and perilous way down to fetch help. It was properly dark by the time the alarm was raised.

I fell asleep to the sound of rain, but as the forecast had indicated, when I woke in the pre-dawn darkness, the rain had stopped. Pat and I wasted no time in getting airborne and flying over the Kaimais. It was still dark as we got to Te Puke, so we had to circle until there was sufficient visibility to approach Gibraltar Rock. The police put several cars with their lights on in the paddock at the foot of the rock, a tall, scrub-covered protuberance in the midst of neatly laid-out orchards and farmland. That helped us to pinpoint the location. The sun was just rising in the east as we reached the scene. There were lots of vehicles—emergency services and media—and the figure of a woman pacing to and fro on the verandah of a nearby house beneath us as we descended.

As we wound the engine down, a cop approached. There was no urgency any more, he told us. His friends had managed to get some pain medication to him around midnight, but Mike Avard had lost too much blood and had died around 3am.

'Who's she?' I asked, nodding in the direction of the woman who was still pacing backwards and forwards near the house.

'That's the wife,' the cop said. He was just a young bloke.

'Does she know?' I asked.

'No,' the cop replied. 'We haven't told her yet.'

That was often the way in those days. Everyone always tried to get someone else to take responsibility for that kind of job.

'Do you mind if I have a word to her?'

'Nah. Go for your life.'

I told Pat to get the gear ready, and I walked over to where Mike Avard's wife was pacing like a caged tiger. She glanced at me as I approached, but kept right on walking.

'Hi,' I said. 'My name's John. I'm flying the helicopter. Can I talk to you?'

'If you can keep up,' she said, still walking. I fell into step beside her. Her arms were folded across her chest and her shoulders were hunched, as though she was cold. I guessed she was in her late 20s or early 30s.

'What's your name?' I asked.

'Christine,' she replied. 'It's my husband up there. Mike.'

'Do you have any kids?'

'Three,' she said.

'Where are they?'

'Back home in Otumoetai. Someone's looking after them.'

'Do you know what's happened?' I asked.

She still didn't look at me.

'No one's told me anything officially,' she said, 'but I've overheard the radios. I've got a pretty good idea Mike's dead.'

'Unfortunately, that's the case,' I said.

She looked at me then, a swift, brave glance.

'Thank you for telling me that,' she said, and kept striding through the wet grass.

I told her that we'd done everything we could to get to the scene the previous night but that the weather had beaten us. I told her that if we'd managed to reach him, we probably could have saved him. But the weather had beaten us all.

'What we're going to do, as soon as it's properly light, is get up there and put him on a stretcher and winch him down. There is a lot of media coverage of all this going on. You need to think about your children. It'll be reported on the radio, so you probably need to be at home with them when they find out what's happened.'

This was true, but I also had something of an ulterior motive in suggesting she left the scene. The poor bugger had been caught in a rockslide and had lain there, bleeding heavily, all night in the rain. He wasn't going to be a pretty sight, and I wanted to spare his wife the sight, if I could.

But she shook her head.

'No, I want to be here,' she said, and shrugged. 'I need to see for myself that it's true. The kids'll be fine.'

It was soon daylight. Pat and I lowered the cliff rescue team onto the narrow ledge, and they secured the body to a stretcher. We winched it up and got Mike Avard aboard. He was a mess, all right.

When we landed, Christine Avard started to walk over.

'You go talk to her,' Pat said. 'I'll try and clean him up a bit. Give us five minutes.'

I intercepted Christine and told her that the paramedic wanted to wipe her husband's face and put a smile on it before she saw him. She understood. We waited until Pat gave the signal and then walked over. She looked at her husband's face, bent and gave him a hug, and then straightened up.

'Goodbye, Mike,' she said.

The police took over from there. Who knows what they said or did. They were all pretty young, and had no real training in these kinds of situations. There was no such thing as victim support back then, any more than there was trauma counselling for those on the front line of emergencies. Pat and I packed up and returned to Taupo along the Kaimais, which were bathed in sunshine.

Christine Avard used to drop in and see us when she came through Taupo from time to time, bringing her kids with her. I was always grateful for the opportunity to see her, and to watch the children as they grew up. I have lately lost touch with her, but I seem to recall hearing that she recently remarried.

***

They were always the worst, those cases, where for whatever reason, usually the weather, you just couldn't get the job done. It was pretty rare for us to reach a patient who was still alive and have them expire on the way to hospital. Of course, it did happen from time to time. In January 1987, we were called to a horrific crash on the Desert Road, where a man had apparently suffered a heart attack while driving, causing his car to drift across the centreline and collide head-on with a van driven by a 35-year-old Auckland woman, Daphne Rangiahua. Daphne's husband and four of her five children were lucky, and suffered only minor injuries. But she was trapped in the wreckage of her vehicle, and while the ambulance crew tried hard to stabilise her, the fire service used cutting equipment to try to free her, which took the best part of 90 minutes. We had arrived and set down on the road where Pat and the local ambulance crew worked to try and keep her alive, intubating her and administering drugs. It's a terrible feeling, knowing you are powerless to help as time ticks by and the patient's chances slowly ebb.

There are few feelings worse, but one of them is the realisation, which overtook us as we headed towards Palmerston North with Daphne aboard, that the weather wasn't going to let us through. We had to turn and make a dash for Taupo instead, but even this was touch and go. I had to open the door of the helicopter so that I could keep the Lake Taupo foreshore in view as we crept towards the hospital. But I could tell from Pat's frantic activity that we were losing her. In fact, by the time we arrived at Taupo and the emergency crew met us, she was already lost. Pat would never give up, no matter how futile the cause seemed.

The weather was a constant factor in mountain rescues. Things had to be pretty bad before we would consider it too dangerous to fly, but we had our moments. In the middle of winter in 1987, we were called up to the Tukino skifield to attend a woman who had broken her lower leg while skiing. It was a complete whiteout, so while we could follow the access road up to the field, we were flying blind from the road-end up to the ski patrol hut where she was being cared for. In the end, I had to get poor old Pat to jump out and flounder along through the waist-high drift ahead of me so that I could use the dim view of his orange high-vis jacket as a point of visual reference. Even so, we couldn't get all the way there, and they had to load the poor woman into a four-wheel drive and crawl half a kilometre down to us as a bulldozer cleared a path. It was touch and go, but by extracting her by air, we saved the victim a rugged, seven-hour journey by road vehicle. The terrible irony of this rescue was that the victim, Debbie Emery, was killed skiing at Tukino just four years later. When word reached Tongariro National Park Headquarters that an avalanche had occurred, a large, coordinated rescue effort ensued. It was late in the day and the light was failing, but I managed to get up to the field with a specialist search team before a blizzard set in. Six of the party of eight were pulled from the jumbled mass of snow and ice alive but injured; Debbie Emery and another Aucklander, John Munro, took 25 and 45 minutes longer to find, and while there was a big effort to resuscitate them, it was too late. Meanwhile, the visibility had closed in and ice was accumulating on the rotor blades. I was pretty keen to get out of there.

This accident happened nearly a year on from one of the worst tragedies on the mountain. In August 1990, two instructors and 11 young army and navy personnel on an outdoor course were overtaken by atrocious weather on Ruapehu's summit plateau. They had dug three snow caves, but had retreated to only one of them when some of the party started showing signs of hypothermia. As the blizzard dumped 3 metres of snow on top of their cave, they began to fear the roof would collapse, and it was decided to try to make a dash for the Dome Shelter, only a few hundred metres away. High winds and furiously driving snow meant they couldn't make it to the Dome and nor could they retreat to their snow cave. They were forced to try to improvise what shelter they could, but couldn't dig more than a few inches into the snow before they struck ice. As two of the stronger members of the party set out to try to fetch help, the others huddled in this scrape in the hillside.

The pair who went for help managed to raise the alarm after an epic, 11-hour slog through some of the worst weather anyone could remember on the mountain. A rescue team was assembled, but there was nothing to be done. It's tough, knowing that you have the technology to bring aid in a matter of minutes to people trapped in an awful situation like this but are quite unable to use it. Decisions are made by the entire team, which stops anyone getting carried away with the stupid impulse to be a hero. You might lead the discussion, telling the other people around the table what you can do and what you'd like to try. In this case, I suggested that even though it was impossible to approach the area where the men were pinned down from the Whakapapa side, it might just be possible to get to the summit from the Turoa side, which might then offer a chance to mount a rescue or at least to deliver survival gear.

We did give it a go, but it was no use. The top of the mountain was firmly in the grip of a fierce blizzard, and it wouldn't let us in. When finally rescuers reached the party, they found that six of the young men had died. The remaining seven were more or less badly affected by hypothermia and frostbite. In many ways, it was a bit of a miracle that anyone survived at all.

***

This abortive rescue was at the back of my mind when we got a call late one evening towards the end of 1991 to try to pick up a 26-year-old man who was suffering from hypothermia on the Tongariro Crossing. He had been tramping in company with a pair of German tourists when the weather deteriorated, and they had attempted to return to the hut at Ketetahi from which they had set out. It became plain he wasn't going to make it, so his companions pitched a tent for him and went for help. Monique Koornz, a climbing instructor with a group of school children who were also on the mountain, went to his aid.

Pat and I had a go at reaching them at midnight in their exposed position near the Blue Lake, but the weather was simply too bad. After returning to Turangi to wait for a break in the weather, we had another go just before sunrise. Again, we were beaten back by fog and high winds. We were able to drop a rescue team off at the snow line, but I was beginning to fear the worst. But later that morning, the ground party reached the group and found the victim in sufficiently good shape to get him to the hut, and we flew him down off the mountain as the cloud base lifted.

I was called in to fly his rescuer, Monique Koornz, off the mountain, too. I congratulated her on getting him through the night.

'I just did what I had to do to keep him warm,' she said modestly.

I knew what that meant. The recommended method of treating hypothermia in the field is skin-to-skin contact—climb into a sleeping bag with the victim and get close.

She was a pretty young woman, blonde and athletic, and it wasn't hard for me to imagine the line of questioning she would face when she met the media contingent who were staking out the rescue coordination centre in Turangi.

'You know what they're going to ask you, don't they?' I said to her.

She looked at me, and I saw the comprehension dawn in her eyes.

'Thanks for the warning,' she nodded.

'How did you know when he was warm enough to make it?' I asked.

'Oh, I knew, all right,' she smiled.

***

Occasionally, the search and rescue business delivers up surprises. In December 1986, we got a call to Acacia Bay, just across the bay from Taupo itself, where an 11-year-old boy had been pulled from the water after spending some time submerged. Pat had responded with a road ambulance, and as he neared he could see a group of people gathered around a slight, pale form on the volcanic sand of the beach. They were administering cardio-pulmonary resuscitation (CPR), and I had a bad feeling about his chances, given he hadn't responded after all this time.

Pat was pessimistic, too, when he heard the details. Adam Cox had set off to swim to the diving platform moored about 25 metres off the beach. He didn't make it, but it was some time—his distraught dad, who was there and who had dragged him from the water, reckoned it was 12 minutes—before he was found. When his dad finally found him, he was unconscious on the bottom of the lake a good 5 metres down. It took the father several goes at getting down to him before he was able to pull his unresponsive son to the surface, a superhuman effort in its own right. Luckily for Adam, there were two firemen and a doctor on the beach, and they immediately set to work doing CPR.

Even so, it looked like Adam was going to be just another entry in the long list of Lake Taupo drowning victims when Pat, who had taken charge of the resuscitation effort, announced that he had detected a heartbeat on the monitor.

Adam was loaded in an ambulance and taken to Taupo Hospital, where they did what they could to stabilise him. Then Pat and I flew him to Waikato, where he was put in what they call a drug-induced coma—heavy sedation, which allows the brain to recover from trauma, and in Adam's case, his lungs to recover from being filled with lake water. As we flew back to Taupo from Waikato Hospital, Pat was pessimistic about the boy's chances of survival: he reckoned there was only a 10 per cent chance he would see out the day. He reckoned it would take a miracle for him to pull through.

But we had no time to dwell on Adam's prospects. We hadn't even reached Taupo when we were called to Ruapehu to pick up a kid who had broken his leg in a fall. We took him to Taupo, and then we were called across to Whakatane to transfer another 11-year-old who had suffered serious head injuries in a bike accident to Waikato Hospital. On the way, we were asked to go and assist with a fire that had broken out on Mount Maunganui. We had to turn that one down.

I caught Rowan Hyde's eye as our latest patient was wheeled into intensive care.

'Back again?' he said.

'How's that kid we brought in earlier?' I asked.

'He's hanging in there,' Dr Hyde said. 'It'll be a couple of days before we'll know.'

A day or so later, they decided Adam was capable of breathing without assistance. And a day or so after that, they withdrew the drugs that were keeping him unconscious. Incredibly, he woke up a few hours afterward and while he couldn't remember anything about the events that had landed him in hospital, he showed no signs of impairment. The cold water and the timely administration of CPR had done the trick.

Perhaps the age of miracles wasn't yet past.

***

Barry Shepherd got hold of me one afternoon and told me that the pilot of a glider involved in the national gliding championship had gone missing somewhere along its triangular route over South Waikato. I collected Barry and Pat and we flew off to see if we could find the missing craft.

It had been flying from Te Hoe further north to Ohaaki about 25 kilometres north-east of Taupo, where it was expected to change course for the final leg to Matamata. The organisers thought it would be somewhere in the vicinity of the corner at Ohaaki.

So it proved. As we buzzed a pine plantation in that area, Barry said: 'There.'

We could see a parachute canopy snarled in the branches, the crumpled remains of a sailplane hanging in a tree, and as I nosed toward it, we could also see what looked like a body suspended from the branches lower down. There was no sign of movement.

'Doesn't look good,' Barry said.

We set down in a paddock downhill from the site and walked unhurriedly up through the trees. I was wondering how the hell we were going to get the body down.

As we got close to the tree, a voice called out: 'About bloody time!'

We all jumped.

'You're supposed to be dead!' I yelled up to Bill Walker, the pilot, who was very much alive. It was a delicate operation getting him down. He turned out to be pretty knocked around, but his worst injury was a broken pelvis. He told us he had clipped the top of one pine tree and crashed into another. He was knocked out, and in a fair bit of pain when he came to, but managed to climb down the tree and spend the night on the ground. In the morning, he was thirsty, and remembered that he had a bottle of water in the cockpit of his aircraft, where there was also a radio. He climbed back up, tried the radio—with no luck—and had the presence of mind to spread out the parachute. Then he slipped and got wedged in the tree, and after a fair bit of straining and wriggling, discovered he was quite unable to free himself. That's how we found him.

We winched Pat into the trees and he cut Bill out of his parachute harness and strapped him into a winching harness instead. We lifted him out of the trees, leaving Pat perched on a branch.

We went back for Pat, but I took care to place the hook a few feet away from him, just out of his reach. He was thinking of taking up a full-time, fully paid rescue position, which would mean he would no longer be available to volunteer for us.

'So what's it to be, Wynne?' I asked him over the radio. 'Are you staying or leaving?'

'Bastards,' he said. 'Suppose I'll stay, then.'

We pulled him out of the tree and he stayed on as a volunteer for many years afterward.

***

I was usually able to keep an emotional distance between myself and the search and rescue work. The exception was when children were involved. I always found those jobs the hardest.

One evening, we were called to a farm near Te Kuiti where three kids were being driven around the property on the transport tray of a tractor by their granddad. When he went to drive up a steep hill, the tractor tipped backwards. Two of the kids fell off, but one of them was holding onto the rail of the tray and his hand was caught between the cab and the metal bar he was holding. His granddad ran off to fetch help. It was late afternoon by the time we reached them. The fire service was on their way, but the poor little bugger had already been trapped for three-quarters of an hour. Rather than make another flight to the road to fetch the fire fighters and their cutting equipment, I grabbed a hacksaw from the tractor tool box and began cutting through the steel while Pat administered pain relief. It took a good 15 minutes, and the boy was remarkably calm throughout. Once the bar was through, we were able to bend it and free him.

It's afterwards that the shock of it hits you. Once we were back at base, I felt it. I suppose it's having kids yourself that makes you particularly susceptible to the pain and fear that children suffer in an emergency. But I got a lot of satisfaction out of this rescue. The boy's life was never in danger, but for once, instead of just being the ambulance driver, I felt as though I had actually done something.

***

It's easy, when you work for yourself, to forget how to say the word 'no' to your customers. It's an attitude that can get you into some funny situations. That's what I was thinking as I put down the phone a little after 4.30pm one Friday afternoon in April 1992.

In 1988, Tony Quayle, head of the New Zealand Meteorological Service, had got in touch with me to ask whether I thought it would be feasible to provide helicopter emergency response cover to their remote weather stations on the Kermadec Islands, 1000 kilometres north-east of Cape Reinga, and to Campbell Island, 681 kilometres south of Invercargill. I said yes. I didn't really give it much thought, because I thought the chances of being called upon to fly out that far were pretty remote. But I talked it over in general terms with Grant Biel. He calculated the amount of fuel you would need to fly direct from Invercargill, if the wind wasn't too strong, and reckoned that with some kind of long-range fuel rig you could probably do it. As far as I could tell, if we had sufficient fuel, the greatest challenge in operating at this range would be navigation. There was no comprehensive GPS coverage in those days, so you'd be relying on the methods used by your crusty old nineteenth-century sea dog, dead reckoning—plotting speed and compass course on a chart, making due allowance for wind speed and direction. It's not as simple as it sounds in theory, but Grant reckoned you could overcome that difficulty, too. Like your crusty old nineteenth-century sea dog, Grant had an extraordinary ability to gauge wind speed and direction by looking at the sea, which came from years of experience in delivering Fletcher aircraft to various destinations from Hamilton where they were built.

None of us really thought we'd ever have to make a mercy dash that far out. But in 1990, the Met Service began to investigate using our helicopter to assist with a resupply and change of the crew of the Campbell Island base. A fuel dump was established in 1991 at Enderby Island, the northernmost of the Auckland Islands group, as a waypoint. Research indicated that the last time an aircraft had landed at Campbell, it was a Catalina flying boat, which made the trip to uplift an injured Met Service worker in 1951. The plane had a bit of a nightmare landing in Perseverance Harbour, a narrow waterway between steep hills aligned with the prevailing westerly and prone to vicious gusts of wind. The plane was damaged and couldn't make the return flight until it had been repaired. The injured man was obliged to wait until a ship arrived for his ticket back to New Zealand after all, so you could say that there had never been a successful aerial rescue from Campbell.

Well, that was promising!

We sought and obtained special permission from the Department of Conservation to place a fuel dump on the largest of The Snares Islands, a tiny cluster of seabird-infested islands about 200 kilometres south of Invercargill. That would give anyone flying to Enderby Island in the Aucklands group a bit of a margin in case bad weather beat them back. Then on 24 October, Grant and I flew to Invercargill, which was already new territory for me: I hadn't previously known that New Zealand went that far south. And the next day, with Grant aboard, we flew over the sullen Southern Ocean south of Stewart Island to The Snares, where we landed and placed a big rubber bladder on the ground and transferred fuel into it from the helicopter. We returned to Invercargill, topped up the chopper and loaded extra fuel to enable us to fly to Enderby Island where we would meet the supply ship *Daniel Solander* and carry ashore an automatic weather station, which was specially designed to provide sheltered storage for drums of jet fuel for future use. We set off again for Enderby via The Snares. Enderby Island was a damp, muddy rock rearing from the sea in those days, the swell gnawing at the fearsome northern cliffs and sea lions littering the beach at Sandy Bay, above which we landed. Many years ago, they had attempted to farm Enderby, but to no good effect. The Department of Conservation was just now in the process of eradicating the last of the introduced mammals from the island—goats, rabbits, tiny mice and a few head of lonely cattle. But thanks to all those gnawing and grazing critters, there was little vegetation, apart from the dense groves of tortured rata at either end of the island. Once we had finished flying fuel and other equipment ashore, we topped up the chopper again and then headed even further south toward Campbell Island, some 165 nautical miles (305 kilometres) away.

There had been few times in my life when I had been completely out of sight of land, and I wasn't sure I liked the feeling. Worst of all, I knew what was involved in putting together a long-range rescue at sea; the best qualified person to plan and execute a rescue if we fell out of the sky out here was Grant Biel, who was sitting beside me! That wasn't a particularly comforting thought, should something go wrong.

But after an hour and a half of staring out at nothing very much, a dim smudge on the horizon indicated that our arrival at Campbell Island was imminent. It grew closer, a rampart of cliffs with raging breakers at their feet. We flew in over the shark-fin-shaped Dent Island, swept over the scrubby slope above North-West Bay and crested the ridge, from where golden tussock receded to the long, Prussian blue finger of Perseverance Harbour. The Met Service station there was a cluster of buildings in Beeman Cove on the harbour's northern shore.

I landed next to the boatshed by the wharf, and we had a cuppa with the five crew, whose stint in that isolated place was coming to an end: the Met Service were changing the guard, and I ferried the relief team and their gear ashore from the *Daniel Solander,* which had steamed there overnight and was at anchor out in the harbour. We also brought a year's supply of food, diesel and other essentials from ship to shore. The wind was blowing a steady 40 knots with the occasionally much stronger gust, which made lowering the empty cargo net attached to the hook at the end of the 120-foot line into the ship's hold interesting, to say the least.

We overnighted at Campbell, and the next day we overflew one of the albatross colonies to take photographs, to test an idea the Department of Conservation had that aerial photography would be a more efficient method of counting nesting birds than walking about amongst them on foot.

Then, after refuelling, we set off from Campbell Island for the return flight. We flew back via Enderby Island, refuelled and then completed the return journey to the mainland via The Snares. By the time we set down back at Invercargill, we had probably offered a great deal of reassurance to the hardy souls who worked way out there on Campbell. They now knew that if the worst should happen, help was only a matter of a few hours away, all being equal.

None of us suspected we'd have to do it for real, let alone how soon.

***

On Friday—it's always a Friday—24 April 1992 at about 4.30pm, I got that call from Tony Quayle.

'Mike Fraser has been hit by a shark,' he said. 'It sounds bad. Can you go and get him? We need to fly him back for urgent medical attention.'

'Yeah,' I said. 'No problem.'

Then I hung up the phone and sat there wondering how the hell we would manage it.

It wasn't great timing. My Squirrel helicopter had only 10 hours to go before it would need a maintenance inspection. This had been scheduled for the following week.

I started making calls. The first one was to Grant Biel, to ask for his help. He agreed at once to come along. The second was to the engineers to tell them not to go home just yet, as we needed them to build and fit a long-range fuel system that would double our range from 425 (700 kilometres) to 850 nautical miles, a whisker over 1500 kilometres. Since Grant was going to be in the cockpit with me, I needed David Brown to fly Grant's high-altitude-capable Piper Navajo overhead.

New Zealand didn't have full 24-hour GPS service south of Invercargill in those days: at best, the coverage was intermittent, with one-hour windows occurring in the coverage twice a day. The Piper was fitted with an Omega navigation system that used VLF (very low frequency) radio signals transmitted by surface stations in six nations worldwide, and didn't require satellites. And, of course, I wanted Pat Wynne.

I got another call, this time from the Wellington Search and Rescue Centre. I told them I was already putting it all together. They said they would do what they could to assist us as we flew south.

Tony gave me as many of the details as he knew. Communications between Campbell and the mainland were by high frequency (HF) radio, and there was no direct link to the site where the victim was being tended to at North-West Bay. We were relying on second-hand information and everything was initially a bit sketchy. All we really knew was that Mike was in need of urgent assistance.

***

Just after lunch, four of the Met Service staff—three blokes and a woman—and the Department of Conservation's research officer, Jacinda Amey, had set off from Beeman Cove to go body-surfing at North-West Bay. They took snorkelling gear in case there was no swell: it's a one-and-a-half-hour walk over to North-West Bay, so they didn't want to walk all that way for nothing.

And sure enough, once they got there, they found the sea to be flat. So they got their gear on and went snorkelling instead.

Mike Fraser, the Met Service's officer in command of the Campbell Island station, was about 30 metres offshore in about 5 metres of water. He was a very experienced diver—a qualified SCUBA instructor, in fact—and he was completely at ease in the water. He was enjoying himself amongst the kelp, looking for juvenile sea lions to play with, when he felt a tremendous blow from his right. He was momentarily underwater. His immediate thought was that a bull sea lion had given him a nudge, but when he resurfaced, he found himself staring at the tonsils of a huge shark. He yelled and went to strike at its nose and its eyes, but he found his left hand unresponsive and his right forearm dangling uselessly from the elbow.

He kept flailing at the animal with whatever was left of both arms, and kicking and headbutting—anything to try to keep it off him. Suddenly, it backed away. Mike's theory afterward was that his right forearm detached and gave the shark something to chew on. Other people who know about sharks reckon it's typical for a great white to make an initial, slashing hit to mortally wound its prey and then to retreat to allow it to bleed to death.

That's what started happening with Mike. He lay on his back and began kicking toward the shore, but the water around him was dark with his blood and the shock was beginning to set in. He would never have made it on his own.

The others had seen everything. Like Mike, their initial thought when they heard him yell and saw the splash was that he had been surprised by a sea lion. The juveniles especially get their kicks racing up to you with their jaws agape, swinging away at the last second. But Mike kept yelling and they saw the dorsal fin of a shark—a bloody big shark—and they knew at once what was happening.

Jacinda Amey, the DoC ranger and researcher, didn't hesitate. She swam to Mike and caught him in a surf lifesaver's rescue hold. He jerked when he felt her hand grasp him under the chin: he thought his head was about to be bitten off, but Jacinda's calm voice said: 'Kick.' They kicked together back to shore, and a wave washed them onto a rocky platform. The group immediately began to do what they could to staunch the flow of blood from his wounds. One of the Met Service workers, Linda Summersby, had spent three days in the Auckland Hospital Accident and Emergency Department learning a few tricks of the trade before her deployment out to Campbell. She'd never dealt with arm amputation before and had no equipment, but she improvised a dressing for the stump of Mike's severed arm with a plastic bag and insulation tape. More tape wrapped tightly around Mike's wetsuit acted as a tourniquet. Meanwhile, Gus McAllister, the island mechanic, set off at a run back to Perseverance Harbour to raise the alarm back in Wellington. He covered the distance that had taken them 90 minutes earlier that day in just half an hour.

Back at Beeman Cove, Gus swept his arm across the shelves of the medical cupboard, dumping everything in a pack and headed off towards North-West Bay. Meanwhile, they radioed both the mainland and the Ministry of Agriculture and Fisheries vessel (as she was then) *Tangaroa,* which was three hours from Campbell. *Tangaroa* immediately began to steam toward the island.

***

The engineers had done their thing and the Squirrel was good to go. We had built and fitted the long-range fuel system that I had dreamed up—three 200-litre drums of A1 jet fuel lashed in the back of the cabin, with an electrically operated pump and associated plumbing that supplied fuel to the main tank via a fitting drilled into the filler cap. We hoped it would work. The trouble was, with the drums in the back of the chopper, there was bugger-all room for Pat. The best plan was to minimise the amount of time he spent threaded around and through the fuel drums and the plumbing: he would fly to Invercargill with Dave, and we would pick him up from there when we stopped to refuel just after midnight.

We were constantly getting updates from the Met Service on the weather. Conditions for the flight were good, and just kept on getting better—to the point where I wondered whether they were talking them up just to encourage us to get going to bring their guy home.

The Piper arrived in Taupo with Grant. Dave and Pat then got in the air ahead of us and set a course for Invercargill. On the way, Pat got a call from an Invercargill doctor who announced that he would be going south with us instead of Pat.

'Fine,' Pat said, but then asked him whether he would be okay with the procedure for performing in-flight refuelling, and whether he was aware he'd be kneeling the whole three hours each way. He also asked the man whether he was confident he could deal with the kind of trauma he was going to treat, because there was no medical back-up way out there.

There was a long pause on the end of the phone.

'Good luck,' the guy said, and hung up.

We lifted off at around 8.30pm, making for Invercargill via Wellington and Christchurch, a distance of a little over 600 nautical miles, which would give our long-range fuel system its first operational test. A magic carpet ride had been set up for us all the way to Invercargill, with other air traffic delayed or diverted in order to give us uninterrupted transit. This saved us time, and time was of the essence.

As we got abeam Christchurch, I said to Grant: 'It works.'

'Looks like it,' he said. 'Don't think we'll need to plan on refuelling on Enderby. We should be able to go direct to Campbell from Invercargill.'

'As long as the weather guys have got it right,' I said.

'As long as they're right,' he nodded.

We were getting a raft of forecasts, each loaded with detail, as we went. I asked them to simplify it for me. 'Is it good, or bad?'

'It's about as good as it gets,' they said.

We arrived at Invercargill around 1am. While we set about filling up every fuel tank and drum in the helicopter, Pat, who had ransacked the hospital, was trying to stuff medical gear and supplies into the cabin. He had managed to get Mike's medical file and had located four units of his blood type. We put both front seats as far forward as they could go to give Pat the maximum possible amount of room, but he was still going to be crammed awkwardly amongst the drums, and Grant and I would have our knees around our ears. Grant is a big boy. It was going to be an uncomfortable bloody flight all round. But I was more concerned about weight. I started shaking my head as Pat went to put yet another box or bag inside.

Sure enough, we were way too heavy for the departure from Invercargill. There was no question of vertical take-off. To get off the ground at all, I had to get translational lift, which is generated by forward momentum, so we sledged along the tarmac with the titanium friction pads on the underside of the skids setting up a shower of sparks. This caused a mild sensation on the ground. Everyone who had been watching us packing the machine to the gunwales legged it for cover.

***

As darkness fell at North-West Bay, the decision was made to move Mike off the beach and up to a point closer to the little sea-lion observatory hut on a bluff above the bay. Once the adrenaline wore off, he was in tremendous pain, especially when they moved him. Amongst the stuff that Gus had brought back from Beeman Cove was a little vial of pethedine, a strong pain reliever. But the label stated that it was not to be administered without medical approval. The gang radioed *Tangaroa* and asked them to seek medical approval and advice on the dosage.

The call from *Tangaroa* was initially received at the Rescue Coordination Centre in Wellington. Then it was relayed to Invercargill and then to the appropriate medical personnel at Southland Hospital. A doctor then gave approval and recommended a dose of 5 millilitres. Back through the chain it went, but in the process, the 5 millilitres became 25 millilitres.

No sooner had they injected Mike with the drug than someone found a guide to the correct dosages of pethedine, and they realised their mistake. Luckily, they had injected the drug into the muscle rather than directly into Mike's veins. But even so, they took turns talking to him to keep him awake, reckoning it would be tragic to save him from the shark only to kill him with a drug overdose. Mike didn't seem to mind. By now, he was feeling no pain whatsoever.

In the early evening, the *Tangaroa* dropped anchor out in the bay and a team came ashore to assist. We now had direct communications with the site of the emergency using the marine channels. A tent was pitched to keep Mike as warm as possible, and everyone prepared for a long night. They had controlled the bleeding as best they could, but they knew it would be touch and go whether Mike would still be with them in the morning.

***

So here we are, with the humming whine of the turbine in our headsets and nothing whatsoever to look at out our windows. The Piper is overhead at 22,000 feet. When Dave relays the wind speed and direction, Grant uses his vast experience and his knack for it to predict what conditions up there will translate to down here at 8000 feet, where we're cruising.

After ninety minutes' flying, we've drained a drum of fuel. It's Pat's job to change the drums over.

'Do we need the empty?' he says. 'We could make more room if we chuck it out.'

'Yeah, dump it, I reckon,' I reply.

There is a muffled curse and sounds of discomfort from the back. In changing the standpipe from the empty drum to a fresh one, Pat has managed to soak the crotch of his flight suit with A1 jet fuel, which has set his reproductive equipment burning.

'I'll slow down and yaw the machine to the left. You get your feet against the drum, and when we open the door, kick it out the left side,' I tell him.

I lower our airspeed as far as I dare. The door opens, and Pat shoves the drum out into the darkness with both feet. We slam the door, and I bring us up to speed again. Pat can now sit on the floor, which is an improvement, but the whole cabin reeks of kerosene.

'Shit, that stings!' Pat mutters.

There is a worrying amount of cloud evident as we begin to close on the point where we expect to be able to see the island and it is still dark outside even though it is close to 6am. A nagging dread begins to gnaw away in the pit of my stomach that we might have come all this way only to be repelled by Southern Ocean sea fog. We don't have enough fuel aboard to give us the luxury of waiting for a break.

After a blank spot of a couple of hundred kilometres where there was no satellite coverage, our GPS suddenly comes alive again.

'I must be losing my touch,' Grant says, comparing his plot with the GPS coordinates. 'We're two miles right of track.'

After this bit of excitement, we're back to waiting again.

By the time the sky ahead begins to show the faintest signs of daylight, all three of us in the chopper have been awake for 24 hours. But I don't feel tired, and I'm sure the others feel the same. The knowledge that Mike's life depends on us keeps us going. Above us, Dave turns the Piper back to Invercargill. He doesn't have enough fuel to travel all the way to Campbell and back. He's had a long day, too. When he got our call to assist, he had just landed in Auckland fresh from delivering an organ to Christchurch Hospital for transplant surgery. He's definitely done his quota of good deeds for the day.

We should be seeing the island any minute now. There is plenty of cloud below us.

I call the skipper of the *Tangaroa* and ask what conditions are like. He confirms that there is total cloud cover. He says he will let go a flare or two and turn his floodlights on to give us a visual fix on their position.

'Got you,' I tell the *Tangaroa.* 'We're getting pretty close now. How's the patient?'

'Still with us, as far as I know,' he reports.

It's about 6.30 in the morning, but it's still pitch black, as it is in those latitudes at that time of year. We're about a mile out, and we begin looking for a way down. In our pre-brief before descending from our cruising altitude, we decide that if we haven't got a glimpse of the ocean by the time we've reached 500 feet, we'll bail out and head back to Enderby Island to refuel and try again. I turn us away from land and let us slowly down through the murk, watching the instruments to ensure we're level while Grant and Pat look for any sign of the earth's surface.

It's getting tense as we near 1000 feet. It's starting to look like, Enderby, here we come. Then Pat says: 'I can see whitecaps.'

I turn us back towards land.

I can't see a bloody thing except the dim gleam of whitecaps, and I know Dent Island is lying between us and land.

'Can you turn your lights on?' I ask *Tangaroa.*

'They're on,' they confirm.

'I can't see you,' I say.

'We've got you on radar,' they reply.

'Can you see Dent Island? Do you reckon you can talk us around it?'

'Yes, sure.'

The helpful operator begins giving me instructions: Dent Island is bearing X, turn Y degrees to port and so on, until I say: 'I'm not particularly bright. I've been awake and flying for 24 hours. You'll have to make it real easy for me. Tell me to go left or right as I head towards land.'

That's how we do it. Soon we're safely around Dent Island and we can see torchlight ashore.

You would think Pat would have learned by now. It's important to stay hydrated, but those of us who spend a lot of time in aircraft know that it's better to take lots of little sips than great big swigs. Pat's been swigging away since we left Invercargill, and now he's desperate to get out. The first I know about this is when I've brought the machine to a hover near the place where the tent is pitched and I hear the door open. The tussock beneath us has been flattened by the downwash from the rotors and looks like a new-mown lawn: I've been here before, and I know better. But before I can warn Pat, he has clambered onto the skid and jumped. He disappears completely from view.

I back off and turn the searchlight on, but we can't see any sign of him.

A few minutes later, Pat calls us on his handheld radio.

'Can you come and pick me up?'

'Where are you?' I ask.

'No idea,' Pat says. 'I feel like Alice in bloody Wonderland down here.' He has discovered that the sea lions have worn a maze of deep corridors in the peaty soil between the crowns of tussock down there.

Eventually, we see his hand poke through the thrashing grass. It's like the opening scene from a zombie movie. He grabs a skid and hauls himself aboard.

We drop 50 metres back to the beach and set down, pushing another empty drum out of the cabin.

We make our way up to where we can see the dim figures of people clustered around the tent, their eyes gleaming in the landing lights as they look towards us.

I recognise Mike at once. He's pretty hard to miss. He's a big, strong bloke with long, curly hair and a bushy moustache. I remember him from when we restocked the island just six months before. He's alive and semi-conscious as Pat examines him. He's waxy with blood loss and a pethedine overdose. His mates have done an incredible job of controlling the bleeding and keeping him from going into shock, and you have to admire the courage of the man himself, too. We load Mike into the helicopter with Pat and fly them both back to the base at Beeman Cove. While I do another flight to North-West Bay to pick up the rest of the team, Pat establishes an IV line and begins tipping fluids into him.

By the time I've returned, he's done a thorough examination of Mike's wounds. He's not happy with the colour of the left hand, which is white.

'I need to operate on that to improve the blood supply,' he says.

I'm nervous about this idea.

'Christ, Pat. We're miles from anywhere and we haven't got any more blood. If he starts bleeding...'

'If I don't do something, he'll lose that hand. He'll be completely armless, as they say.'

So I act as anaesthetist, or at least, as an anaesthetist on a mediaeval battlefield would have acted, holding Mike down while Pat works his magic in the mangled tissue of his forearm. To get at it, he has to unwrap a dressing complete with a snorkel that the team used for a splint. Then he gets to work. It's a success, and you can actually see Mike's fingertips go pink. Only then does Pat excuse himself to go and have a shower and tend to his privates, which have been marinating in jet fuel all this time.

Refuelling is a challenge. Gus brings a few drums of fuel down on the tractor, but while

I'm transferring fuel, there's a noise. A creature not unlike a wet, grossly obese Jersey bull flops heavily toward me. It bares its long, yellow teeth and grunts at me, and I'm hit by a gust of its foul breath. I don't know much about the New Zealand sea lion, and my pilot training hasn't equipped me for this situation at all. I remove the nozzle from the fuel filler pipe and squirt some kerosene at the animal, which grunts in surprise, turns, and flops back into the water again. But it's only temporary. Either the same one, or another one, is back a couple of minutes later. So that's the Campbell Island fuel routine: pump, pump, arf! arf! squirt, splash, pump, pump, arf! arf!...

After I've finished refuelling, Grant uses the base radio to call the mainland and activate our return flight plan. We lift off with an expected flight time of three and three-quarter hours. I ask them to direct Dave Brown to meet us halfway.

Then all hands carry Mike out to the chopper on a duvet: there was no room for a stretcher, of course. He's fully conscious by now, and his colour has improved. We wedge the poor chap in as gently as we can between the two fuel drums and the front seats. We have to ask Mike to sit across the helicopter, partly upright and with his legs bent so that he will fit.

'Teach you to be such a big bastard,' I tell him. Despite the pain and the discomfort which is etched all over his face, he grins.

Pat clambers in after him and makes himself as comfortable as he can. Grant and I climb in and we ready ourselves for departure. The group of people stand and watch as we rise into the air. Among them I can see the diminutive figure of Jacinda Amey.

I've got mixed feelings. I'm elated that we have Mike aboard and alive, because there's a good chance he's going to survive. But I feel sorry for his team, who we're leaving behind. The whole thing—the shock, the fear, the gnawing anxiety, the sheer physical exertion—will catch up with them. They're going to be feeling a bloody long way from home.

***

The flight home is pretty much routine. As we're nearing the south coast of New Zealand, I ask Mike where his family is.

'Napier,' he replies. 'My mum's there.'

'Let's give your mum a call,' I say. He gives me her number, and I dial it on my cell phone. After a couple of rings, it's answered.

'Mrs Fraser,' I say. 'John Funnell here. I've got Mike in the chopper with me here—he is alive and okay. Would you like a word with him?'

I reach behind me with the phone. No one takes it. After a second or two, Mike pipes up.

'Funnell, you dopey bugger. I haven't got any arms!'

Pat holds the phone to Mike's ear so that he can reassure his mum. He tells her he'll soon be on the ground.

***

Invercargill airport is thronging with cars and people.

'Shit, what's going on?' I say to Grant. 'Must be an air show or something.'

There are police, fire and ambulance vehicles on the tarmac, and heaps of people both at the perimeter fence and airside, many of them with cameras.

'Oh, God,' I say. 'Is there somewhere else we can land?'

But we set down. People rush forward, and there's a throng around us as Mike is unloaded from the chopper and handed over to an ambulance crew. We're buggered. There's no question of trying to fly on from here. I had made arrangements for one of our pilots, Pete Masters, to come and fly the helicopter home in the morning. Pat and I will go back with Grant Biel and Dave Brown in the fixed-wing that afternoon.

After the crowd has cleared, a policeman approaches me. I've never quite managed to watch a policeman approach without a surge of panic and a quick search of my conscience for any information on what he's likely to want to talk about.

'Mr Funnell?' he says.

I have to admit that I am.

He reaches out his hand, and introduces himself as a senior sergeant with the Southland Police.

'I've heard a bit about you,' he says. 'But quite frankly, we didn't think you'd pull this off. We didn't think you'd make it. Well done.'

I shake his hand and mumble that it was a team effort.

'Now, you know where to send the bill, don't you?' he says, producing a policeman's notebook. He flips it open, scribbles a post office box number on a piece of paper, tears it off and hands it to me. I tuck it into the breast pocket of my flight suit.

'Out of curiosity, what do you think it'll all cost?' he asks.

I do a few mental calculations. Twenty hours at approximately $1500 per hour. And then there's the fixed-wing.

'Grant? What do you reckon the fixed-wing would have cost?'

'Ooh,' he rubs his chin. 'No, don't worry about it. If I calculated at mate's rates, something like $1000 per flight hour...'

'Yeah, so without being too precise, I reckon a ball-park figure would be $45,000,' I say.

The cop's eyes bulge.

'Forty-five...'

Words fail him. Suddenly, he snatches the piece of paper from my pocket and legs it, vaulting the airport security fence and disappearing, never to be seen again.

Only then does it occur to me that I have no idea how we're going to pay for the whole show. The Philips Search and Rescue Trust has an annual budget of $35,000, and I have blown it on one job. It's hard to see them seeing the funny side, and I begin to suspect they will feel I have well and truly exceeded my financial authority.

When we get back to Taupo that afternoon, 37 hours after getting the call from Tony, I talk it over with Trish, and she has the excellent idea of approaching the Accident Compensation Corporation.

She phones the ACC office in Napier, and they're encouraging. 'Send it in,' they say.

A couple of days later, I get a phone call. 'Forty-five thousand? You'd better direct this claim to Wellington.'

Meanwhile, of course, we're getting messages of congratulations from all over the show, from the Prime Minister to the Secretary for Transport, Dame Margaret Baizley, and the media are all over us. They're interested to learn that we are making a claim for recompense to the ACC. Gavin Roberts, the head of the corporation, can't ignore the general public feeling that we don't deserve to be out of pocket for the mission. We receive the money within two days of the claim being lodged in Wellington.

***

Not everyone is so easily impressed. The Department of Conservation aren't too thrilled to learn that we littered the pristine Southern Ocean with a kerosene drum. And a week after my return to Taupo, I get a phone call. It's the Civil Aviation Authority.

'How did you get your helicopter to Campbell Island, Mr Funnell?'

'Um, flew it.'

'Yes [sounding pained], but how did you keep it flying? What were your fuel arrangements?'

'Drums of fuel in the back seat.'

'Was that approved?'

'Well, I approved it.'

'You know what I mean. Did you get CAA approval for that?'

I ask him whether he thinks there'd have been any chance, if I'd phoned to ask permission to build an unproven, improvised fuel system so that I could set off immediately on the longest rescue mission flown by a single-engined helicopter, they would have given me approval, just like that, let alone on a Friday afternoon.

They were less than impressed, so for the next two to three weeks, I occasionally wonder when the official axe is going to fall. I even phone a journalist friend of mine, Kingsley Field, a reporter for *The New Zealand Herald,* and tell him I think the CAA are going to throw the book at me.

Kingsley can already see the headlines, and he phones back the next day to tell me that the paper is being inundated with calls from law firms eager to take my case for free.

Still there is no word from Civil Aviation, so Kingsley phones them. They tell him that due to the public interest in the matter, they have decided not to proceed with a prosecution. But they also tell him—as they vow to me—that they will be watching me closely.

The ZK-IWI lands at l'Esperance Rock.

CHAPTER NINE

PEAK PERFORMANCE

We did a rash of long-range rescues in the early 1990s. In May, just two months after the Campbell Island dash, we were called upon to recover the body of a drowning victim at Raoul Island. Grant and I took the Hamilton machine this time, fitted with two long-range tanks that gave us a range of 1000 kilometres. We planned to refuel at Great Barrier Island and l'Esperance Rock—the southernmost outlier of the Kermadecs, and not a large piece of real estate, where Grant had established a fuel dump as part of a resupply mission in 1990—but it proved to be unnecessary, thanks to a tailwind. The victim in this case was a 50-year-old woman who had been SCUBA diving off her yacht with her husband. The police were initially quite suspicious of the circumstances surrounding this death: the victim's husband elected to sail their yacht to Whangarei while we carried the body. But eventually they satisfied themselves that there was nothing untoward about it. The wind, which had been our friend on the outward flight, was against us on the return trip, so we had to land at l'Esperance, after all.

A year later, it looked like they needed us to go back to Raoul, as a yacht was in difficulties well to the south of the Kermadecs. The *Windsong,* an American yacht crewed by its owner and his wife, had rolled and been dismasted. The woman had suffered facial injuries, and medical attention was required. We put together a plan that would have had us fly to Raoul, refuel there and then fly south to a container ship stationed halfway, refuel again, fly to the boat and return via the same, complicated route. That mission was cancelled, and a frigate was sent instead. But then, in June 1994, we were called back to Raoul to evacuate a Department of Conservation worker who had a case of iritis, an acute infection of the eye. We weren't so blessed with the weather this time, and from the moment we took off from Ardmore aerodrome in the dark, we faced strong headwinds and rain en route. The workers on the island had to light our landing with the headlamps of the island's vehicles.

In September 1992, six months after the Mike Fraser rescue, we winched a Russian seaman who was having heart problems off his trawler on the Bounty Rise, well to the south of the Chatham Islands. And then, in mid-July 1993, we became the first helicopter to fly to the Chatham Islands themselves.

***

Flying to the Chathams wasn't in itself too difficult. The trouble was the subject of the rescue was well east of the Chathams, in a trawler way the hell out on the Chatham Rise. He had suffered a series of epileptic fits, and his condition was causing enough concern that the skipper requested helicopter assistance. At that point, they were 1000 nautical miles (1852 kilometres) east of New Zealand, which was simply impractical. But between the skipper and myself, we worked out that if he steamed toward Chatham Island, by the time we had got organised and flown out to refuel on the Chathams, he would just about be in range. Using the winch to get people aboard is possible in the dark, but it's far from ideal: it's far better to do it in daylight. What's more, the airport on the Chathams isn't set up for night flying. I figured out that if we left New Zealand at daybreak, we would have time after refuelling on the Chathams to get out to the ship and back in daylight. My rule of thumb was that I should have an hour's worth of fuel in reserve in any given situation. Grant's calculations indicated it would be just possible to work at that margin if all the timings were kept to.

The skipper and I agreed that he would phone me with an update on his vessel's position and the condition of the patient at about 3.30 the following morning.

I got to work on the phone to put together a team to plan and execute the rescue. Grant

Biel had a meeting the next day in Wellington, but he was happy to send his apologies for that and come along. Warren McKay would accompany us in a Piper Chieftain with long-range fuel tanks. Pat Wynne would be the paramedic. Dean Yeoman, who was one of our engineers and a pilot, and Marty McNamara, who was a helicopter pilot herself, would come along and crew.

When the boat's skipper phoned on his satphone as agreed, he reported that the patient's condition seemed to have improved. He hadn't suffered a seizure for a few hours, so it now seemed unnecessary to lift him off.

With some relief, I phoned around everyone and stood them down. At about five in the morning, I went back to bed and fell asleep.

I woke to the sound of the phone. I sort of knew I was going to hear the tinny sound of the satphone before I even had the receiver to my ear, and sure enough, it was the trawler's skipper again.

'He's had another horrendous fit,' he told me. 'You need to come and get him ASAP.'

'Christ,' I muttered, looking at the time. Everything had been premised on our departing Taupo at daybreak. 'We'll see what we can do. How fast can that thing go?'

He agreed to steam as fast as he could to reduce the distance I'd have to cover by air. He told me he thought the patient would be okay to put into a harness and be winched off the deck, so that was the plan.

I phoned around the team again. I caught Grant just as he was about to board a commercial flight to Wellington for his meeting. He agreed to renew his apologies, and our plan was that Warren would fly to Auckland to collect Grant and they would then head to Napier to rendezvous with the helicopter. Then I phoned Dean Yeoman, who was already at work at the hangar.

'Start the machine,' I told him. 'We're off out there, after all.'

The long-range fuel arrangement was already installed and full. With Dean doing the pre-flight and starting her up, it was literally a matter of my getting to the airport with Pat Wynne and jumping in. We were three hours behind the schedule I'd worked out for the original plan, which made me nervous. But, luckily, the wind speed and direction was in our favour, so I was able to tell Warren that we were going to flag our planned fuel stop in Napier and head directly out to Chatham Island, where we would rendezvous.

We made good time, mostly because Grant was able to get Warren to hunt up and down to find favourable winds for us. And, because the fixed-wing reached the Chathams ahead of me, they were able to jack up the fuelling arrangements, so that as soon as I set down they were ready to start pumping.

Then, after loading Grant as co-pilot, we set off to cover the 300-odd nautical miles (555 kilometres) to the trawler, which was steaming steadily toward us.

'How's fishing?' I asked the skipper, to make a bit of idle conversation.

'Not too bad,' he said. 'We're basically full and we would have been on our way in anyway. We've done pretty well.'

'What're you catching?'

'Bluenose,' he replied. 'You want some?'

I looked at Grant. He looked at me.

'Sure,' I said.

Again, Grant used his uncanny knack of finding favourable winds to give us a tailwind out there. We still had a bit of daylight left by the time the trawler hove into view.

'How's the fuel?' I asked Grant. He told me. I worked out that we must have a nearly full drum of fuel that we'd planned on burning and hadn't needed. It seemed to me that it would be surplus to requirements, and we could certainly afford to lose a bit of weight. But when I suggested we jettison the drum, Grant was dead against it. The return flight was going to be into the wind.

So as we prepared to do the winching, we were heavy, and my engine torque gauges were showing we were uncomfortably close to the limit of our available power when in the hover. The skipper was offering to heave his boat to, but I instructed him to steam directly into the wind as fast he could make his vessel go. I wanted to keep the helicopter in translational lift, as the technical term has it: a helicopter uses less power when moving forward than it does at the hover. Soon we were positioned a little ahead of him and moving at 15 knots, which was delivering us a bit of a bonus, as we weren't pulling a lot of power. We would need all of that margin, because we were about to add another person to the load.

Pat began to operate the winch, and the line streamed behind us as it lowered toward the trawler. I was eyeing all my instruments.

'They've got the line,' Pat told me. 'They're just hooking the guy up.'

There was a bit of a pause.

'Yep. They're giving us the thumbs-up,' he reported.

'Okay. I'll take the weight,' I said.

It's prudent to test a lift before you commit yourself to it, especially when you've got a real, live human being on the end of it. If you find the machine's performance is too badly compromised, your only option is to place the load back on the deck.

Grant, as previously mentioned, is quite a big man, and fortunately he was sitting on my right. As soon as I took the strain on the winch line, the chopper adopted a distinct list to port. Under ordinary circumstances, I would have eased back on the collective and lowered the load again to reconsider my options. But Pat said: 'He's clear of the deck.'

The voltage indicator was hovering close to the red line as Pat operated the winch. I increased our forward speed to try to reduce our engine power demand.

'He must weigh a ton!' I said desperately.

'He's got a pack on his back and he's hanging onto a bloody big box!' Pat reported.

'Fish!' I realised. 'I thought he meant he'd give us a fish or so, not a whole box!'

With the downwash and our forward speed, the poor patient was being blown this way and that, with the box acting like an aircraft wing. It seemed to take forever to get him up to the door. I was terrified we would burn the winch motor out, in which case we would have no option but to drop the guy into the sea for the boat to recover. But eventually he appeared in the doorway, box and all.

'Grab the fish!' I yelled to Pat. Once our bluenose was secure, Pat bundled the patient—a young guy, in his twenties, I would guess—into the cabin and closed the door. The last light was ebbing from the sky, and we still had a couple of hours to run to get back to Chatham Island. Somehow, Grant and Warren organised us another tailwind for the return journey, too!

The locals had parked cars and brought anything portable that would light up to light the runway for Warren and for us, and we were met by the Catholic nuns who ran the little

Chatham Island Hospital. They took over our patient, who was pretty chirpy, and they were happy to keep our fish frozen for us, too. We went to stay at Waitangi Lodge, right on the water in Petre Bay, and the beer tasted pretty good that night. So did the fish, all 25 kilograms of it. We were eating it for months.

***

Closer to home, there was no real let-up in the rescue work, either. By now, the Trust Bank helicopter in Hamilton was dedicated full-time to air ambulance work, somewhat relieving the load on our machine, which continued to be a working helicopter that could be used for search and rescue or ambulance duties as required. There was no shortage of work. In late 1991, the Philips Search and Rescue Trust extended its service to the Manawatu, with a Squirrel dedicated to rescue and ambulance work based at Palmerston North Hospital. We bought this from Grant and Billie Biel, who made it available to us at a very advantageous price because they continued to be staunch believers in what we were doing. It performed its first rescue the day before the service was officially launched: pilot Steve Oliver was called to fetch a forestry worker who had been injured by a rolling log.

But soon after we launched the Manawatu service, Philips New Zealand contacted us and told us that while they were still committed to funding the trust, the electronics industry had suffered in recent years, and they were no longer in a position to support us at the level they had been doing. Their decision was a bitter blow, but it was hardly a surprise. Half of our funding came from agencies such as ACC and the police, but without sponsorship, we couldn't continue. We had minor sponsors, mostly the major employers in the region—Tasman Forestry and Electricorp, along with Shell Oil and Telecom—but we needed a new major sponsor if we were to continue with our work. We had done 700 rescues since the establishment of the trust, and were doing them at the rate of one and a half a week.

We didn't have to wait too long before a replacement for Philips came forward. I met Francis Small from New Zealand Rail, and by the end of March 1992 we were able to confirm that the Taupo and Palmerston North machines would be known as the Tranz Rail rescue helicopters.

***

In March 1995, we celebrated the tenth anniversary of the setting up of the Philips Search and Rescue Trust. Plenty had changed. In that first year, we had done 32 search or rescue missions. I had mostly been a commercial helicopter pilot with a sideline in SAR. In our tenth year of operations, by contrast, the Taupo-based machine had flown 152 missions. Partly the increase reflected a reduction in the capability of the outlying hospitals, which had been scaled down in the successive waves of restructuring of the health sector in the late eighties and early nineties. Patients, who might once have been treated by the 'stay and play' brigade, were now 'load and go' to Waikato. Partly, it was because our operation was more capable. Besides better and better medical gear, our search and rescue equipment had become more sophisticated, too. We had better communications gear—radios, pagers and even cell phones, now, even if cell phones were still awaiting miniaturisation. The searchlight on the helicopter, night vision goggles, winches and then socalled FLIR (forward-looking infrared) imaging equipment meant that we were one of the better set up SAR operations in the country. That meant more call-outs. We were running three helicopters, now, and plans were well advanced to replace the Hamilton Squirrel with a twin-engined IFR Bell 222B. I was now the chief executive of a large search and rescue operation, who moonlighted as a commercial helicopter pilot. Trish was a fulltime administrator—and was also the person who was usually on the end of the telephone when a call came in. Even Mark and Hamish were involved, helping to clean and polish the aircraft and assist with fuelling and loading medical gear.

It was fun looking back on the years we'd been in the business. We had scores of cuttings and lots of letters from people grateful for the assistance we'd rendered them in their hour of need. One of my favourite examples of these was this one:

> *I would like yous to accept this donation of $100, for helping me during my accident of 29/11/89.*
>
> *I know its a long time since, but I've always had yous all in my mind.*
>
> *Thank you all so much.*

It wasn't the $100 that touched us, although I have little doubt it was a lot of money for the writer to part with. It was the sentiment. It came straight from the heart, and that's what it was all about. We had helped make a difference in someone's life.

***

Some things about search and rescue didn't change. We were still kept busy by the roads, the farms, the forests, the rivers, the lake and, of course, the mountains, throughout the 1990s. The high, snow-capped volcanoes of the Central Plateau had long been the scene of some of the more technically challenging and dangerous flying missions I was called upon to perform. Back when I started at Taupo with Helicopters New Zealand, the Lama was the only turbine-engined machine in the North Island.

Unlike petrol-powered engines, the Lama's jet turbines were less troubled by the thin air at altitude, which meant we were much in demand for rescues from the alpine reaches of the Tongariro National Park. With Don Bogey, and before we had winches, I pioneered a method of extracting people from hillsides using a long line. With this, you would swing alpine rescue people wearing crampons and carrying ice axes to the slope adjacent to the victim. They would unhook, secure themselves and then prepare the victim for lifting. Once all was ready, I would swing them to a location where it was safe to set them down, land and load them aboard. Don subsequently moved to the South Island and this system went with him. It was widely adopted, and is still much used today. It is faster than the winch, and it minimises the time victim and rescuers spend being blasted in the freezing downdraft of the rotors.

On one body recovery operation back in those days, I had brought a RARO (Ruapehu Alpine Rescue Organisation) team to a steep pitch on Ruapehu, where the spring ice conditions had caught out an unprepared climber, as they commonly did. I had landed on a small ridge. Everything was so icy that I had to wriggle the machine left and right to dig the skids in. I was using the Lama on this occasion, and it was good for this type of work because you could stop the rotors but leave the engine running.

The team clambered out, geared up and set off into the gloom towards where the body of the climber was lying. I sat for a while, but decided after some time that I needed to have a leak. I was dressed warmly enough, but I tended not to wear crampons in the machine and discouraged others from wearing them, too: they punched holes in the floor. I climbed out. I planted one foot to test the footing, then gingerly planted the other. As soon as I committed my weight, my feet slipped out from under me and I was on my way. I just managed to hook my hand around the cross-member to the skid. Had I not managed to do that, the next stop was the bottom of the precipitous gully, three or four hundred feet down. The rescue team would have been momentarily puzzled when they arrived back at the chopper to find it unattended, but it would have been a simple matter of following the blood stripe on the icy rocks. Strangely, I no longer felt any interest in toileting duties.

In October 1993, I was called to one of the more distressing jobs I had done. A boy who was tobogganing on a plastic bag went over a bluff on Whakapapa skifield. Another boy, perhaps his brother, tried to grab him, but in all, three of them ended up falling 10 metres onto rocks and then variously sliding and tumbling another 150 metres before coming to rest at the bottom of Skippers Canyon. We were on the scene with a rescue team within 20 minutes of getting the call, but by the time we were able to begin lifting the boys out, a 14-year-old had died. His brother, who was 15, died en route to Waikato Hospital.

In 1995, a team of geologists chartered me and the chopper to fly them to Mount Ruapehu's crater lake, where they wanted to do some monitoring. I had done this a number of times, and knew the lake and its vicinity pretty well.

Ruapehu's biggest eruption was in 1945, and its most notorious activity was on Christmas Eve in 1953, when it produced the lahar that washed out a rail bridge at Tangiwai, killing 151 people. It had erupted once or twice since then, but had been otherwise pretty much dormant. On the way up, the scientists told me that the chemistry of the lake had changed subtly in the last little while, and there was some anxiety that the mountain was entering an active phase. I already knew this, because a TV crew had got wind of the scientists' mission and had asked me to fly them up so they could film them at work. I couldn't do it, because I had no room. So they had given me a little camera and asked me to do some shooting instead.

I set us down on a little cinder beach on the edge of the lake, and the scientists spread out with their equipment. I didn't like to shut the machine down when I was up there, so I left it running with the doors open as I stood and filmed them doing what they do. As usual, gusts of steam were blowing off the milky surface of the lake. I had just lowered the camera to wipe the lens when I heard all this shouting. I looked up. I saw the dim figures of the scientists all charging away from the lake—and a wall of water about a metre high sweeping towards me.

I ran to the chopper and without even pausing to put my headset or seatbelt on, I increased the engine RPM to flight idle and lifted off. The beach beneath me disappeared underwater. The scientists had all found spots on the icy area around the lake to perch out of harm's way, and they were all frantically beckoning me back. By clamping the cyclic between my knees, I managed to get my headset on, and could then speak to them over the radio. I directed them to a spot at a safe distance from the crater where I picked them up. No one was in a mood to hang around.

A few weeks later, on the afternoon of Saturday, 23 September 1995, the mountain erupted in earnest, throwing a spectacular cloud of steam and ash heavenward and spilling a couple of small lahars toward Whakapapa. The mountain was evacuated, and I flew a few sweeps of the slopes to make sure everyone had been safely brought down. Two days later, I was called back up to the airstrip near the Chateau, where a light plane that had been chartered by a group of scientists hadn't quite made it as far as the runway. None of them were seriously injured, but I took a couple to Waikato Hospital as a precaution. And I also got a call from a freelance TV cameraman, Geoff Mackley, who had sneaked past a Department of Conservation cordon in the early hours of the morning and tramped up to the higher slopes to film the activity at close range, which is what he does and why I know him as 'Mad' Mackley.

'It's getting pretty hairy up here,' he said. 'There's rocks and stuff falling all around me. Can you come and pick me up?'

Rocks! You'd have to have rocks in your head! But whereas anyone else would walk down the hill away from the danger, Geoff does the opposite. Wherever the action is, you'll find him there, and he has a well-deserved reputation for getting footage no one else would dare to get.

***

A few months after the eruption had quietened down, in October 1995, an operation had begun to spread sodium fluoroacetate—better known as 1080—on the bush around Tongariro National Park. There's nothing like mentioning the use of this poison to get people worked up, whether for or against. It's a nasty substance, and it causes quite a bit of distress to animals, especially dogs. Deerstalkers hate it, as do people who advocate for a possum-trapping industry. But I knew from my time doing surveys of tuberculosis in the Hawke's Bay wild deer population that without taking some kind of action to control deer and possum numbers, we could kiss our agricultural industry goodbye. There was nothing as effective as 1080. I had also seen the effect on Raoul Island, which had significantly reduced birdlife and tracts of ravaged vegetation when I first visited, and had been transformed into a lush, green place full of the song of birds that had repopulated from adjacent, pest-free islands. That was the work of 1080. So all things considered, and until a better solution was devised, I couldn't see an alternative to its use.

One of those who would never quite get around to seeing it that way was Chris Short, who used to work for me part-time. One day, as Alan Beck, the Taranaki helicopter pilot who was flying the poison in, was loading, Chris appeared dressed in camouflage gear, carrying a pack on his back, a rifle in his hands and several others of various calibres slung over his shoulders. He didn't point the rifle at anyone, and he didn't issue any threats, but it was plain he wasn't there to congratulate them on the good work they were doing with 1080. He instructed Alan and the loader-driver to get into the big Bell chopper and told them to fly him up the mountain.

There is an emergency code that you can send from aircraft operating in New Zealand airspace that alerts traffic control that you have been hijacked. Alan activated the code and it was received, but the authorities decided it must have been a false alarm.

When they reached the upper slopes of Mount Tongariro, Short directed the pilot to land. He jumped out of the helicopter with his pack and his guns and walked off into the rocks. The pilot wasted no time in getting into the air and getting the mountain between him and Chris. He contacted the police and told them he had been hijacked. And in the meantime, all hell broke loose, with the media getting hold of the story. That, of course, was exactly what Chris wanted.

The police considered their options. The spot Chris had chosen was well protected with the rocks providing plenty of cover and a panoramic view of the open, steep slopes surrounding his position. A tactical approach was out of the question. Chris was known to be a master bushman and an expert marksman. Better leave him to it, for now, they decided.

Meanwhile, Chris's two brothers were anxious to see the whole thing end peacefully. One of them walked up and had a yarn to Chris. They then talked to Noel Christiansen, another bloke who used to work for me and who was a mate of Chris's. Noel got in touch with me.

'His brothers are pretty keen for you to go in and bring him out,' he said.

I wasn't overly keen. Chris was a heavily armed and passionate man who had just hijacked a helicopter. It wasn't exactly a dream job.

I consulted a few other people, including Barry Shepherd. He consulted other cops, and came back to me.

'We can't tell you not to do it,' Barry said. 'We reckon don't, but it's up to you.'

Thinking about it, I decided that since Chris knew me, and so far as I knew we were on good terms, I could probably risk it. So I agreed to go.

Chris's brother came along for the ride. We located him reasonably easily. He was just sitting there, with his rifles laid out in plain view on the ground. I landed beside him, and his brother went to talk to him. When he came back, he told me that Chris was willing to come down. He had made his point.

We made a deal, whereby his guns went into the locker and he agreed to come with me. Chris didn't make any show of resistance whatsoever, and was perfectly calm and friendly. En route, we were able to tell the police that he wasn't going to put up any kind of struggle. Just before he landed, he remembered he was still carrying another weapon: it was a difficult thing to conceal—a samurai sword that he had forged himself out of the leaf spring of a car and to which he had fitted a handle ornately carved from a stag's antler. He didn't want the police to confiscate it. I agreed to look after it for him. We landed, he was quietly taken into custody and the whole episode ended much better than was expected.

Chris spent time in jail for the hijacking, and when he got out, he came and saw me and told me that I could keep the sword, as a gift for flying him off the mountain. A few years later, he was diagnosed with terminal cancer. He managed another peaceful protest on the slopes of Mount Tongariro, but he didn't last that much longer. I still have the sword. It is a beautiful object, and razor sharp.

***

A month after we marked the anniversary of the trust, we performed a rescue that sort of typified how we had come to do things, and what we were now capable of. At about 7.30 one dark night in April 1995, I got a call from the Hawke's Bay police operator to say that they had an emergency at Timahanga Station in the Kaweka Range, close to the summit of the Napier–Taupo Road. A hunter had been shot, and because the local rescue helicopter lacked night vision gear, they couldn't do anything to help. Pete Masters and I mounted the FLIR to our helicopter and got ready to head off.

It so happened there was a search and rescue training night at the Taupo police headquarters, so rounding up the usual suspects was easier than usual. Pat Wynne was there, as were Barry Shepherd, Tony Roddan and Wayne Murtagh, all very experienced SAR hands and willing to go at a moment's notice. They arrived promptly, and we got under way at around eight.

We flew to a woolshed at the southern end of Timahanga Station. There was a very tired and frightened looking nine-year-old boy. Anthony Pol had been hunting with his dad, Henry. It was the roar—breeding season—the most popular time for deerstalkers to head for the hills. Somehow, Henry had been shot through the knee. Anthony had done his best to splint it, but it was too painful for Henry to walk on. He instructed Anthony on how to get back to the hut from which they had set out. Anthony was pretty sure he'd be able to find his way back to civilisation from there.

By the time Anthony had found the hut and then made his way down to where he thought the farm was, the afternoon was waning. He had been running and walking through the bush for 10 kilometres. He saw some men working on a fence in the distance. Getting to them required him to ford a swift little river, and he was terrified that the men would have knocked off before he found them. But they were there, and they took him to the woolshed from which they raised the alarm.

The Hawke's Bay helicopter attended, but the light was failing, and Anthony found it impossible to orientate himself from the air. That's when the Napier police called us.

I knew the general area where we were heading, because I had top-dressed Timahanga way back in my Wanganui Aero Work days, working for John Harding out of Taihape. Anthony explained roughly where he thought his dad was, and we loaded up and headed out to look for him.

It was probably his torch that saved Henry Pol's life. It's unlikely we would have seen it with the naked eye, but it showed up on the night vision goggles as a brilliant point of white light. It was windy as hell. There was no way I could approach the steep, bush-clad gully where he was lying, so I flew to the top of the ridge above it.

'I'll drop you up here,' I told the boys. 'I'll fly down and hover over him so you can make your way down.'

I approached the open top of the ridge. It was too windy to try to hover, so I moved forward as slowly as I could. Pat, Barry, Tony and Wayne took turns to climb out onto the skid and jump out into the tussock. Barry gave me the thumbs-up, and I wheeled off to relocate Henry Pol while they got their gear together and started to make their way downhill.

It began to rain.

The terrain was steep and the bush was thick, so even though the vertical distance wasn't great, it took an hour for the four men to reach Henry. When they did, Wayne immediately said: 'Christ! Is that you, Henry?'

'Gidday, Wayne,' said Henry, sounding relieved and exhausted. He had worked with Wayne, and knew him well.

'Got him,' Barry's voice said over the radio.

'We can't winch him out in this weather in the dark,' I told him. 'I'll have to come back in the morning.'

'Roger,' said Barry. 'We'll set up a tent while Pat treats Henry.'

I stayed at the woolshed and spent most of the night looking after Anthony and listening to the rain lashing the roof. I wondered what the injured man's chances were. I was conscious that as afraid as I might be, Anthony would be doing it a thousand times worse, thinking about his dad.

It was, by all accounts, a miserable night. Henry's knee had been pretty much destroyed by a .270 round. He thought another hunter must have shot him. Barry couldn't tell, but now was not the time for asking questions. Pat medicated him for pain, they kept him warm all night and did their best to prevent him slipping into shock by telling what Barry later described as war stories all night.

The following morning was fine. The Hawke's Bay helicopter arrived to offer what assistance it could. I flew in, located the party and we rigged Henry and winched him aboard. Pat was next. Once we had loaded them, I waved Barry, Tony and Wayne goodbye and swung away for the woolshed. The Hawke's Bay machine took my place to winch them out.

Henry's reunion with Anthony was something to see. It was hard to tell who admired whom the most. Henry was flown out to Hastings Hospital by the Lowe Walker chopper. Anthony was a Taupo boy, and he got a ride back with us. It was his second ride in a helicopter, and it's a fair bet he enjoyed this one far more than the first.

The far east of Russia, where I went to inspect and purchase a Mil-8 helicopter.

CHAPTER TEN

AMONGST THIEVES

In 1992, the state-owned enterprise Timberlands West Coast decided it would put the harvesting of its native forestry estate on the West Coast out to tender. Grant Biel phoned me and suggested we form a joint venture company to bid for the work. If we were successful, we would get hold of some big helicopters and make it work. He thought he knew where there were a couple of Mil Mi-8 MTVs—better known as the Mil-8 medium-lift helicopter—available for lease: they would meet Timberlands' requirement that the machines used be capable of lifting a 5-tonne payload.

In the end, we were the bridesmaid in the tender process to an overseas company. Their proposal was pretty similar to ours—they were going to run Mil-8s as well—but differed in certain respects. Civil Aviation were terribly impressed by their operations manuals, and more than once mentioned them in glowing terms to me. If only my company could build and implement manuals like that. It was as though they had set the new CAA benchmark.

A condition of the tender was that a local company did the successful bidder's maintenance controlling, and so they approached us and asked whether we could provide their ground services. We were happy to do that.

However, after a few months we began to hear some unsettling things about the way they were operating that we thought might be creating adverse exposure for us, their flash manuals aside. We told them that we were resigning as their maintenance controller.

A matter of weeks later, on 16 October 1993, one of their Mil-8s went down just south of Hokitika, en route from the aerodrome at Hokitika to the work site, 18 nautical miles distant. All three people aboard—two Kiwi pilots and a Canadian engineer—were killed. No explanation could be found for the crash. However, subsequently our company inspected the wreck at the crash site and found that the friction brakes for the flight controls had been backed off, something that was against the manufacturer's operating procedure. The main rotors had struck and actually severed the machine's own tail and the helicopter had plunged nose-first into the ground.

Timberlands, who found themselves with commitments to various West Coast timber mills, urgently needed an operator to step in and fill the gap. As next best tenderer, we were asked if we would accept the contract on the same terms as in our tender document. Grant and I accepted, and Heli Harvest was born.

***

The two Mil-8 machines that Grant knew of were based in Goroka in Papua New Guinea, and were originally offered to him by a man named Vitali Tchentzov, representing a company named Aerinn based in Moscow. Grant phoned Vitali, and he agreed to honour the original terms that had been negotiated in the run-up to the submission of our tender. Vitali reckoned it was a simple matter of signing a lease, handing over the money and travelling to PNG to uplift the helicopters.

But it turns out that when you're dealing with Russian businessmen—or a certain class of Russian businessman, anyway—there are few certainties. The Papua New Guinea government had seized both the helicopters we were interested in, on the grounds of a tax default. The whole thing looked like a bit of a daunting mess, but Grant undertook to fly to Port Moresby and sort it all out. I was to stand by to travel up and fly one of the machines home if he was successful.

The call came a couple of days before Christmas 1993. Together with two sets of Russian flight and maintenance crew, I flew to Port Moresby—a hot, derelict sort of place, if ever I'd seen one—and then on to Goroka up in the highlands. The crew set to work on Christmas Day preparing and test-flying the helicopters for the delivery flight to New Zealand. When we returned from the test flight, we were met by a police chief from Mount

Hagen, who waved a bunch of papers and told us that we were trying to remove machinery that had been seized by the PNG government. He invited us to board his truck and accompany him to the jail in Mount Hagen while it was all sorted out. It was Christmas, so there was no reason to believe it would be sorted out quickly. No way were we going with him.

Luckily, Grant cuts a pretty imposing figure, and we lined up behind him as he fronted the discussion. He had his own sheaf of papers, and they clearly stated that all the encumbrances on the helicopters had been discharged and that the PNG government had relinquished all claim to them. Trouble was, they were written in English, which the Mount Hagen man plainly didn't speak.

The day got hotter and hotter. I don't handle the heat too well. I was sweating faster than I could tip water into myself. I wasn't relishing the prospect of sitting in a PNG jail cell, let alone in these conditions. But, finally, Grant prevailed on the police chief to get a translator, and he declared himself satisfied that the release documents were genuine. We were free to go.

We loaded both helicopters with as many spares as we could safely carry and got off the ground as quickly as we could, anxious to put the maximum possible distance between ourselves and agents of Mount Hagen law enforcement.

I liked the Mil-8 to fly right from the start. The bigger that helicopters are, the easier they are to fly, and these great, lumbering beasts, weighing in at 13 tonnes, with two turbine engines and a massive main rotor, were much more stable and forgiving than the little things I was used to flying.

We flew to Daru, an island off the southern coast of the main landmass of Papua New Guinea and at the mouth of the Fly River. It was about as far north as Port Moresby, but in a much more direct line with the Australian coast. After clearing PNG customs—a slightly nervous wait—and refuelling there, we flew across the Torres Strait, dotted with islands and coral atolls, and down the east coast of Australia, stopping at one-horse towns on the way for fuel. The Queenslanders eyed the big helicopters incuriously: choppers of all shapes and sizes head to and from PNG all the time, operated by Australian mining and exploration companies.

When we were in the latitude of the Queensland and New South Wales border, we left the Aussie coast behind and cruised out over the Tasman to Lord Howe Island. After refuelling there, we hopped 900 kilometres to Norfolk Island. From there, it was a skip and a jump, relatively speaking, to New Zealand, and we soon had our Russian maintenance teams in Taupo preparing one of the machines for logging operations.

We encountered immediate problems with the regulator. Civil Aviation were concerned that they had lost one foreign-flagged Mil-8 in the last three months already: they were damned if they were going to lose another one, and their oversight was much more hands-on than we would have preferred. It seemed as though they were being deliberately obstructive, to the point where Grant and I attended a meeting in Wellington to discuss what seemed to be a ramping up of their requirements with their acting director. Grant ended up being quite firm and telling them that we were staring down the barrel of a penalty clause if we didn't start honouring our contractual commitments to Timberlands shortly: he advised that we would need to brief the Minister of Forestry if CAA wouldn't allow us to operate.

Within 24 hours, we had been approved to fly.

***

Grant Ward, Wayne Pratt and I took turns as pilot in command doing all the line lifting work. Aerinn supplied Russian flight crews, so we all learned to speak a few words of Russian, and they all learned to speak a few words of English. They were great men, good workers and good fun, too, beneath the poker faces that Russians wear as part of the national costume.

Besides the flight crew, we had a gang of what we called 'hookers', whose job it was to put chokers on the logs—heavy steel cables that tightened to grip the log as the strain was taken up by the helicopter. In some parts of the West Coast, such as in the Hochstetter Forest up in the Buller District, the topsoil was pretty thinly spread over a limestone base, so the trees generally had only a poor grip on the ground. We found that if we pulled collective power while lifting one of the larger logs, the Mil-8 rotor wash could actually blow the trees beneath us down. The answer was to make sure the hookers were well clear before we pulled power. This required a fair bit of planning from the pilot's point of view and considerable physical fitness and sheer balls on the part of the hookers. Once you had located the log you wanted to fly, you would have to approach steadily with the hook dangling 150 feet below. The hooker would have to catch it and hook it to the chokers as you passed overhead. But stuff still got blown down, and the spectacle of some joker sprinting through the forest trying to outrun the falling kamahi trees was entertaining from the air. Not so sure how funny it seemed on the ground!

Grant, who is a bit of a design and engineering genius, was constantly scrutinising our whole operation and looking for ways of improving it. The Mil-8 is famous for being heavy and cumbersome, but Grant identified a whole lot of the Soviet-era communications and navigation gear with which they were equipped as essentially redundant. He prevailed upon Vitali to allow us to strip this stuff out, and that made a huge difference to the efficiency of the machine. He also devised a grapple that replaced the wire-rope chokers as a means of grabbing the logs. We were lifting some massive bits of wood: some of the logs weighed as much as 5 tonnes. The Timberlands guys who were doing the felling got pretty good at estimating the weight, and we had a really efficient operation going there, after a while.

***

My grandfather on Mum's side hailed from the West Coast, but I'd never really had much to do with the place. I loved my time there, and the locals welcomed us with open arms. Not all of them, you understand. Even though Timberlands was engaged in what was called 'selective logging'—pulling carefully selected trees out of the forest in what was supposed to be a sustainable harvest—there was a fair bit of resistance from the environmentalist lobby. There was a group that were following the Timberlands gangs around and climbing or chaining themselves to trees selected for cutting. We were also constantly vigilant against the possibility of sabotage of our machine, one of a helicopter pilot's worst nightmares.

Whenever we were parked in an area where we thought there was a risk, we would engage security to keep an eye on things. But overall, the Coasters couldn't have been more hospitable. By far the majority were happy that we were part of an operation that employed West Coasters in logging and milling the timber, to say nothing of serving food and drinks to the helicopter crews and their ground staff.

The hookers, needless to say, were young, fit and liked a beer at the end of a day's work. I knew most of them pretty well, as we'd brought one team down from Taupo. We were based at the old nurses' home in the mothballed Hokitika Psychiatric Hospital. One day, I woke to start the day and went around the rooms only to find that none of the boys' beds had been slept in. That wasn't good. I had given them permission to head into town the previous evening, but had given them a lecture on behaviour.

'We're guests here,' I told them. 'Respect the locals. And for God's sake, given the choice, make love, not war.'

As I was beginning to wonder what the hell I should do, I heard a car pull up, and three of the boys climbed out, covered in blood. 'Oh, shit,' I said. 'You've got a story for me, I'll bet.'

It turned out that the gang had been sharing a pub with the West Coast rugby team and their hangers-on. One of our boys had taken a shine to the missus of the captain, and the captain took exception. A few verbals rapidly escalated to a brawl, with four of our team against 15 of them. One of our boys tried to even the odds by smashing a glass on a table and using it as a weapon.

That probably accounted for the blood.

By the time the cops had been called and everyone had made a run for it, the pub was a mess and there were various strong candidates for charges ranging from disturbing the peace up to assault with a weapon and grievous bodily harm.

'We didn't want to come back here,' they said. 'We knew the cops would come looking for us here. So we drove into the bush and spent the night in the car.'

I must have looked pretty dark on it all, because one of them said: 'You told us to make love, not war. That's all Norm was trying to do.'

The other two members of the gang, Corby and his brother Johnny, drifted in and we all went to work. We had a Coaster working with us on the ground, Pete, whose nickname was Fat Boy, because he was the master at dodging any real work. He specialised in finding excuses to run errands into town. Some time during the day, Fat Boy disappeared on one of his usual missions. After we knocked off, we heard that the police were looking for Johnny, who was supposed to have been the glass-wielder.

I arranged with Johnny that he would get his gear together and hide in the bushes adjacent to the airstrip at Hokitika. I would taxi my fixed-wing out onto the runway, he would run out and jump in and I'd fly him back to Taupo with me. That's what we did.

Three days later, while I was back in Taupo, the police confronted me and accused me of aiding and abetting a fugitive from justice. I freely told them I had flown Johnny back to Taupo, but I had not been officially informed that he was wanted for anything. So far as I was concerned, I hadn't done anything wrong.

They blustered a bit, but no charges were forthcoming.

A couple of months later, when I was back working the second Mil-8 on the Coast, I got a call from Fat Boy.

'Mr Funnell, you've got to help me. Norm and Corby are gonna sew my lips together with a wool pack needle!'

'Put Norm on,' I told him. 'Norm, what's going on?'

'It's Fat Boy, John. He's the squealer. He got Johnny in trouble.'

'You let him go,' I told him.

'Oh, we weren't going to hurt him bad. Only gonna sew his lips up so he don't squeal again.'

I told them to let the guy be. And they did, for the time being. Two months later, they told the patrons of a pub where they were drinking that if anyone was interested in where Fat Boy was, he was having a snooze in the urinal.

***

The selective harvesting of large trees by helicopter that we were doing for Timberlands West Coast was attracting worldwide attention, and in 1996 we were contracted by a French company to provide similar services in the Solomon Islands. To do it, we decided Heli Harvest needed to purchase another helicopter that would be based in Honiara. We located a brand-new Mil-8 in Moscow and arranged the charter of an Antonov 124 cargo jet to carry it to Kuching in East Malaysia.

The Antonov is a massive aircraft. The only disassembly required to fit a Mil-8 in the cavernous cargo space was the removal of all five main rotor blades and one blade from the tail rotor, both of which were straightforward tasks.

Six days after refitting the rotor blades in Kuching, Grant headed east across Borneo, Northern Indonesia and Papua New Guinea and flew out to the island of Guadalcanal, where he arrived at Henderson Air Base ready to start work in the jungles of Makira Province in the remote east of the Solomons.

Meanwhile, trouble for Heli Harvest had blown up back at home. This time, it was from an unexpected quarter. Aerinn had decided they wanted to alter the terms of the lease of one of the original Mil-8s and charge us 20 per cent more than we had agreed. That's the Russian way, it seems. A contract isn't so much a legally binding agreement as a discussion document. I was incensed, and told Vitali that we were in a contract with Timberlands and, according to the conventional way of doing business in our jurisdiction, we couldn't go unilaterally changing the terms of the agreement.

Vitali couldn't see the sense in that at all. He wouldn't budge in his demands, and since he held all the aces—without his chopper and crew, we had no business—we had no option but to capitulate.

So pissed off was I, though, that I started carefully studying all the documentation relating to the helicopter and its terms of hire. It was in Russian, and I'm no cunning linguist, nor am I that conversant with the Cyrillic alphabet, but I thought I could see something odd.

It so happened there was a Russian family staying in Taupo whom we'd met once or twice. They had a daughter, Marina, who was in her late twenties or early thirties and spoke excellent English. I rang her, and asked if she would have a look at the documents for me. She was happy to help.

'Who owns the helicopter?' I asked.

'It is company called Tyumen Avia-Trans,' she said.

'Not Aerinn?'

'Nyet. Aerinn is leasing this machine.'

I got Marina to talk me through the documents. There didn't seem to be any ambiguity. Vitali didn't appear to be the owner of the helicopter, as he was purporting to be, at all. The owner was a Tyumen-based company.

Talking it over with Grant, we decided we would contact the company's general director, so far as we could tell, a man named Nikolai.

Two days later, Marina phoned Nikolai, and introduced us. He spoke terrible English, about as bad as my Russian.

'Ah, yes,' he said, once the initial hurdles to communication had been overcome. 'I have heard of your friends. They have stolen my helicopter.'

This gave us a bit of a shock. He told us that he had signed a lease agreement with Aerinn in Russia, but Aerinn had never paid any of the fees agreed in the lease while the machine was operated in Papua New Guinea. Worse, Aerinn were now using it to make money by leasing it to us without Avia-Trans seeing a bean of it. I could see his point of view.

'Nikolai, we'd like to help sort this out,' we told him. 'We'll pay for you to fly to New Zealand so we can talk about it.'

Nikolai refused. Vitali was very well connected, he told us, and would soon learn that Nikolai was on his way.

'Well, let's meet in Sydney, then,' we suggested, and after a bit of thought, he agreed.

Grant and I flew to Sydney, taking Marina with us. We met Nikolai in the lobby of the hotel. He was a pleasant man in his sixties, but he was plainly pretty disappointed with us. In the lift on the way up to the conference room, I asked how it was that Nikolai and his company had failed for so long to notice one of their Mil-8s had been pinched.

'Ah, is not new. We have lost six of these machines,' he shrugged.

'Six?' I gasped.

'It is easy when you own so many. We have 600 Mil-8s,' he said.

It turned out that his company was formed to assume the government helicopter business at the collapse of the Soviet Union in 1990. And we thought *we* had hit the big time!

In our meeting, Nikolai was cordial, but plainly disenchanted and inclined to be sceptical that we were innocent parties in the whole thing. But we offered him a deal, whereby we would pay off Tchentzov's crews and Nikolai could step in and take over that role. Meanwhile, we would stop payments on the lease of the machine and enter a lease/purchase agreement with Tyumen Avia-Trans.

Nikolai agreed in principle, but we couldn't seem to agree on the money.

At lunchtime, I said to Marina: 'What's going on? What's the hold-up here?'

She looked at me appraisingly.

'Do you really want this to happen?' she asked. 'I think I can get him to agree, but I would like something in return.'

'What do you want?'

'I would like tickets to Moscow for myself and my brother,' she said.

We resumed the meeting after lunch, and quite quickly, it seemed, we reached agreement, with Marina translating. She produced a set of documents—two in English, two in Russian—and we all signed. We would lease the machine with a right to purchase at the end of the lease. Nikolai was all smiles by now. We all shook hands and departed friends. At the end of the lease term, Grant and I paid for Nikolai to fly to New Zealand and attend a ceremony where we formally handed over the cheque and proudly accepted the ownership documents to the helicopter.

Of course, changing horses in mid-stream as we did wasn't simple. Because we were in effect swapping operators by changing our lease agreement from Aerinn to Tyumen Avia-Trans, we had to clear it all through Civil Aviation, who could have taught the Soviets a thing or two about bureaucracy. But soon we had it all set up.

When the Mil-8 arrived back in Taupo, we called the Russians together and told them that the machine had been seized by its lawful owner, who was going to continue to lease it to us but who no longer required their services. I felt sorry for them. I liked them all, and they were the meat in a strange sort of sandwich here. Being Russian, and being accustomed to this sort of thing, they thought we were pulling some kind of swifty. But when they studied the documents, they were convinced, whereupon they accepted it with a gloomy, collective shrug.

Then we had to break the news to Vitali. Vitali wasn't happy, and he came out all guns blazing. Writs were filed in the High Court in Hamilton, and even though we tried to persuade him to drop it, he pursued the matter for a couple of years, hiring and firing a succession of lawyers and draining himself dry in the process. I've never seen so much energy expended in a lost cause: he reminded me of that cartoon of a frog, halfway down a stork's throat, with its hands clamped tightly about the stork's neck.

***

With one of our machines well and truly bogged down in the ownership dispute in both New Zealand and Russian jurisdictions, we found ourselves in the tricky situation of being contractually obliged to provide logging services both in the Solomons and on the West Coast and with only one machine to do it. We worked January 1997 in the Solomons, February in New Zealand and then flew the helicopter back to the Solomons in March. We couldn't keep on doing it like that. It was essential that we acquired another machine, and we set our man in Moscow, Alexander Kharan, looking. He reported that he had tracked down a Mil-8 MTV Type 8 with less than 30 hours' flight time on it in Khabarovsk, about 800 kilometres north of Vladivostok and close to the border with China. It sounded good, so we entered long-range negotiations by fax with the Vladivostok-based owner. It all proceeded very smoothly, and soon we had agreement on practically every detail of the very detailed deal. Both parties agreed that the chopper was sold, subject to inspection and with the consideration—around a million US dollars—payable when it was free on board a vessel bound for New Zealand.

Grant was happy for me to go and do the inspection, so I arranged to meet Alexander in Vladivostok. I suggested our in-house contract lawyer, Marian Glazewski, accompany us, as I don't rate myself highly when it comes to seeing the fishhooks in contracts.

So, in March 1997, we flew first to Japan, and then on to Vladivostok, the former Soviet naval base city on the Pacific coast of Russia. It was just coming into spring there. There was still ice in the harbour, and the city from above was still blanketed in snow.

Alexander met us at the airport. A short time later, a couple of men in dark suits arrived, too, and they showed us to a black 4WD with black tinted windows. They seemed friendly enough. They took us to our hotel, just across the road from the airport, and then we drove across Vladivostok to where the meeting was to be held. The streets were slushy and pocked with potholes. Every so often, we'd drop into one of these with a crash that would rattle your teeth in your head. One particularly large one made the headlights go on.

We presently arrived at a forbidding-looking building surrounded by a high wire fence topped with razor-wire. We drove in, and were shown into a foyer. We shrugged out of our heavy coats and woolly hats and gloves and were glad of the warmth of the radiators around the walls, but even so, the inside of the windows was thick with ice. What a way to live!

We were shown into a meeting room where we were introduced to the helicopter owner's representative, whose name was also Alexander. Everyone sat down and negotiations opened, and we recorded each item on our laptop and printed them off on a printer we'd brought along. As expected, we made rapid progress through the terms that had been agreed in our chain of faxes. The main hold-up was our inkjet printer, which was playing up in the cold. We found it worked happily if we sat it on the radiator to make the gelid ink melt and flow better. When we stopped for lunch, I said to their Alexander: 'This is going well. We should be done this afternoon.'

He shook his head, a look of profound pessimism on his face.

'Nyet,' he said. 'Will take many days.'

Our hosts took us into town for lunch. The food was beautiful and so, I suppose, was the vodka, if you're into it. There had to be vodka.

A shot glass brimming with vodka was put in front of each of us.

'My friends! I am happy that we do business. Na zdorovye!' Alexander said, and tossed his back.

'Na zdorovye!' we echoed, and drained our own glasses. They were quickly refilled.

'To thank you for your hospitality, and hoping for a fruitful and mutually beneficial relationship!' said our Alexander, who knew the drill. 'Na zdorovye!'

'Na zdorovye!' we echoed, and drank again.

So it went. I slowed down after the first two, but both Alexanders and Marian kept going. Marian was of Polish extraction, and he was therefore constitutionally suited to this kind of thing. I was not.

After lunch, we headed back to the compound. Then, to my utter amazement and complete frustration, everything we had agreed in the morning was suddenly the subject of new and dogged negotiation. By the end of the day, we were actually further from reaching an agreement than we had been at the beginning.

Our delegation retreated to our hotel to regroup. I opened one of the three bottles of Johnny Walker Black Label that I had bought on the way through duty free, knowing how partial to it Marian was. I went easy on it, because while my dad was a whisky drinker and it seems to agree with both my sons, the whisky-drinking gene seems to have skipped my generation.

In the morning, the Russian security team showed up and drove us back to the compound. Even though I had gone easy on the Black Label, I had a headache and was ready for a battle, especially in light of what had happened the previous afternoon. Nevertheless, the morning went quite well. We made good progress, and then we went out for another cordial, vodka-infused lunch.

After lunch, it was exactly as it had been the day before. All our good work seemed to be undone. I was on a short fuse, and after an hour or so of hearing Alexander baldly go back on what he had earlier agreed, I stood up.

'I've had enough of this,' I said to our Alexander. 'Tell him we had an agreement before we even came here. We've been through all the details. These guys are just trying to change it all. Tell him they're not the only people with helicopters for sale in Russia, and I'm buggered if I'm going to do business with them.'

Both Alexanders were talking rapidly in Russian. Marian and I gathered our papers and walked out the door.

In the foyer, there were four big, crew-cut men in dark suits between us and the door. As if to make the point even clearer, one reached inside his jacket and produced a handgun. And in case we were inclined to believe this was fake, he worked the action, and the solid, metal click-clack it made left no one in any doubt that this was real heat he was packing.

'Now what?' said Marian.

'Reckon we sit down,' I said.

A loud conversation was still in progress in the meeting room. We sat in the foyer listening for the best part of an hour and a half before our Alexander emerged, stony-faced. We were driven back to the hotel in silence. When the drivers had gone, Alexander turned to me.

'Don't ever do that to me again,' he said.

I began to tell him that I had no wish to deal with the sort of people they had proven themselves to be, but he cut me off.

'You don't want to mess with these guys,' he said.

Grant Biel phoned the next morning. The vendors' Alexander had faxed him to complain about us, and had told him that we were trying to change the deal. I told him what had gone down.

'Do whatever you think is right,' Grant said.

I told our Alexander that the deal was off, and we were catching an 11.30 flight to Moscow in the morning, which happened to be the next flight out of Vladivostok.

'No,' he said. 'We can make this work.'

'I don't want to deal with these guys. I'm not going back there. They pulled a gun on us.'

'I think I can salvage this.'

I thought about it.

'Okay. You've got till eleven in the morning. If we haven't heard from you, we're on that plane. You tell your wife to expect us at the airport in Moscow. We've got her number, so if you don't call her, we will.'

In the morning, Alexander went off with the two Russians. Marian and I had a leisurely breakfast. Ten came around, then 10.30. We were packed and ready to go, and at 11am, we were just dragging our bags into the lobby ready to walk across to the airport terminal when the doors burst open and Alexander blew in on a gust of cold air.

'It was all a big misunderstanding!' he said. 'They want you to come back and prepare the agreement.'

'Yeah, right,' I said.

'No, they are ready to sign the agreement,' he said. 'Just like before.'

'Well, I'm not going back,' I told him. 'We'll write the agreement here in the hotel and fax it through, page by page, and then they can sign it. If they haven't signed it by eleven tomorrow, then we're gone.'

I offered to help Marian draft the agreement, but he told me all he needed was the laptop, the printer and the second bottle of Black Label, to keep him warm. So I went to bed. In the morning, I went to his room. The door was open, he was asleep, and on the desk was my laptop. I looked at the screen. There was the contract, ready to print, and so far as I could tell, it was a perfect reflection of everything that we had agreed before arriving in Russia.

After breakfast, we began faxing it through. It went off smoothly—until the very last page, the signature page. That simply refused to go. There was nothing for it. We would have to go into the compound again, after all.

Their Alexander, when he met us, was cordial, but plainly disappointed. He seemed to think he had been the victim of some kind of sharp practice. But both parties signed the agreement as Marian had drafted it. Now it was only a matter of inspecting the machine and signing the 'giving and taking' agreement, as the Russians call it.

***

To get to Khabarovsk, we boarded a Russian domestic flight in an old plane. There was no cargo compartment: everyone had to carry their bags aboard and stow them as best they could in the rear of the cramped passenger compartment. The hostess was pretty big in the final drive and had to turn sideways to shuffle up and down the aisle.

After we had landed, I stood up and moved into the aisle ready to disembark. Alexander pulled me back into my seat. Everyone sat there for a minute or two, then the door to the flight deck opened, the pilot, co-pilot and hostess marched off, and only then did everyone else get up and begin to disembark.

'Put your gloves on,' Alexander told me.

'Nah, I'll be right,' I said.

'Put your gloves on,' he repeated. 'And don't touch anything metal with your bare hands.'

Outside, the cold was brutal. My breath froze as soon as it had left my mouth. It was a relief to get into the terminal, where the dominant feature was a cafeteria with huge pots of soup simmering away.

I didn't get to see anything of Khabarovsk, because the chopper was stored at the airport. We were driven out across the tarmac to it, and while we were waiting for the engineers to do their inspection, we took the canvas rotor covers off and sat in the big cabin wrapped up in them. I was only wearing leather-soled shoes, and my toes were killing me.

We did our inspection of the aircraft, went for a flight and it was fine.

So we returned to Vladivostok—balmy by comparison with Khabarovsk—and entered into a new round of negotiations over how to make payment. They were keen on a Russian bank to hold our funds. We weren't. We were keen on a New Zealand bank. They weren't. Both sides clearly expected the other one to try to diddle them.

We suggested a Japanese bank, and after a while they accepted this. They wanted a letter of credit established, and they insisted on a peculiar form of wording. We faxed the agreement to Grant, and he faxed back later that evening to tell us that this would almost certainly cause trouble when they went to draw the funds down. We tried to explain that to the Russians, but they insisted on their wording.

The letter was established and both parties then had to sign it. We were taken into a fancy office with lots of luxurious leather furniture and desks of dark, highly polished wood. Alexander introduced me to the head of the company, a man named Sergei.

Once he and I had signed the many pages of the agreement, two sets in English, two sets in Russian, I produced the last bottle of Black Label, and handed it over as a token of our gratitude for the deal we'd just down. He looked pleased.

'Spasibo,' he said, then yelled: 'Natalya!' A door opened and a gorgeous blonde woman with legs up to the ceiling came in carrying a tray with exquisite crystal glasses on it.

Sergei unscrewed the cap from the bottle and threw it over his shoulder. He slopped a decent slug of whisky in a glass for Marian, for both Alexanders and for me, and lifted his own.

'My friends, it has been our pleasure to do business with you. Happy flying! Na zdorovye!'

'Na zdorovye!'

'Don't be too quick to drink your next one,' I muttered to Marian.

I made a short speech. 'Na zdorovye!'

Sergei tipped his head back to drink. I dumped my whisky into Marian's. Marian knocked his back. I lowered my glass with the others.

So it went, until the bottle was gone. But neither Sergei, who had drunk a quarter of a bottle of whisky, nor Marian, who had drunk only slightly less than half, showed any effect whatsoever. I was the drunkest one there.

***

The saga, however, wasn't quite finished. I made sure I went back a month later to supervise the loading of the Mil-8 onto the Norwegian vessel that was to take it to the Solomons, as I wouldn't have put it past the Russians to substitute something else. But it was our machine, okay. True to form, their Alexander showed up, and from what I could gather, he thought he could somehow renegotiate the contract for more money.

'Nyet,' I said. 'We have an agreement. You have your letter of credit which you can draw down when my helicopter is free on board and the captain has signed the bill of lading. That's it, as far as I'm concerned. If you're going to start trying to renegotiate it, then I walk. This isn't the only helicopter for sale in Russia.'

He gave up. The ship sailed. The Russians discovered they couldn't draw down on the letter of credit, and immediately presumed their fears had been realised. They'd been diddled by these Kiwis.

We were back in New Zealand by now.

'Just tell us the problem,' we said, 'and we'll fix it. We're men of our word.'

The helicopter arrived in Honiara on 18 April 1977 and was put to work shortly after that.

After many days of negotiation, the agreement for the purchase of the Mil-8 is signed.

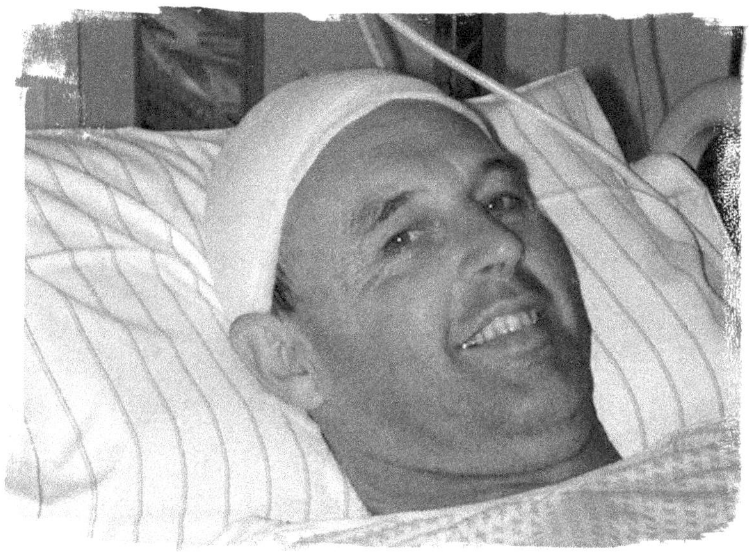

In recovery, 12 hours post-op.

CHAPTER ELEVEN

UNDER THE KNIFE

The world changed for pretty much everyone in September 2001. It had already changed for me.

I had had differential hearing loss—one ear worse affected than the other—for most of my life, but it had never really bothered me much, and nor had I really wondered about it. Aircraft are noisy machines, after all, and I had been around them all my life. I had never had any trouble getting medical clearance to fly (which is mandatory for pilots), until one day in early 2001 I got a letter from the Civil Aviation Authority. A new principal medical officer had been appointed, and she had taken it upon herself to review my file.

I read her letter with a winded feeling.

In her opinion, I no longer met the required standard and in order for her to reconsider issuing my medical certificate, I would need to get a report from an ear-nose-throat specialist as to what was causing the hearing loss.

I had only ever done two jobs in my life: milking cows and flying aircraft. I was buggered if I was going to go back to milking cows. I didn't know what I was going to do.

I did what she had directed me to do, and made an appointment with Wilfred Wong-Toi, an ENT specialist in Rotorua. But in the meantime, I was just plain devastated. I'm normally a pretty positive person. On a scale of one to ten, where one is bad and ten is pretty bloody good, I'm normally a fifteen. I was definitely down the scale at that time.

I went through to Rotorua to see Mr Wong-Toi. He examined me, and referred me to Hamilton for an MRI scan. He didn't say so, but I was pretty sure that meant he was anticipating some kind of brain tumour, most likely an acoustic neuroma, a rare, benign tumour affecting the nerve that conveys signals from your ear to your brain.

There was another wait before my turn in the claustrophobic, white tunnel of the magnetic resonance imaging machine came up. I had seen these done plenty of times after bringing in head injury patients, standing on the other side of the glass where the operators are and watching the pictures come up on the screen.

After the procedure, I wandered through to see if I could get a look at the images. I knew most of the people who ran the machine, but I couldn't seem to find anyone. There was only the office lady around.

'I wouldn't mind seeing the pictures, if I could,' I said to her.

'I'm sorry, but you can't,' she replied. 'They have to be processed.'

I grinned at her.

'No, they don't,' I said. 'They'll be sitting there on the computer now. Someone will let me see them.'

'I'm sorry, Mr Funnell,' she said. 'Everyone's gone.'

That was frustrating. Odd, and frustrating.

Two or three days later, a courier brought the hard copy of the images to my house. I was there with Hugh Barclay, a friend and our in-house solicitor. I held the film up against the window, and there in shades of grey and black was my head in cross-section.

'Is there a brain in there?' Hugh asked.

'Yes,' I said. 'Amazingly enough, there is. But *that* doesn't belong there.'

I pointed to a bright spot on the film at the back of the oval that was my skull.

'That,' I told Hugh, 'is a brain tumour.'

'No,' he scoffed.

'I'm afraid so,' I said. 'I've been with the Waikato intensive care and I've seen plenty of MRIs over the years, and I've heard the operators talking about what's what on the pictures. I know this isn't supposed to be there.'

'It'll be nothing,' Hugh said. 'You should stick to flying helicopters.'

But he looked shaken, all the same.

I tried phoning Mr Wong-Toi's surgery, but his receptionist told me he was in surgery.

I explained that I was an air ambulance pilot, and that a lot of people depended on me.

If I had a condition that was going to impinge on my fitness to fly, I needed to know and so did they.

Grudgingly, the receptionist said she would tell Mr Wong-Toi that I wanted to talk to him. She phoned back soon afterward, but wouldn't discuss anything over the phone. She got me to make an appointment for 5.30pm the following day.

The night and day Trish and I spent waiting was sort of surreal. I had already seen enough to be satisfied that there was something sinister in my brain, but that still left plenty of scope for my thoughts to get ahead of themselves. It was most likely to be benign, but what if it wasn't? And even if it was benign, what if it couldn't be treated? I would still be unfit to fly. And what if it was benign and treatable but something went wrong? I had seen how close the mass lay to my spinal cord ... I'm sure Trish was going through the same iterations.

We drove to Rotorua and sat in Mr Wong-Toi's waiting room. Then we were shown in. My slides were on the lightbox.

'Sit down, Mr Funnell,' he said. 'I have good news and bad news.'

'What's the bad news?' I asked immediately.

'The bad news is that you have a brain tumour. I'll need to refer you to a brain surgeon.'

We never found out what the good news was.

***

In 1999 I had attended a helicopter crash where the pilot was John Dunn, a general surgeon from Auckland. He had told me that if I ever needed his help in anything medical-related, just give him a call. I phoned him and told him that I did need his help. I had been diagnosed with a brain tumour and needed a quick referral to a neurosurgeon.

John jacked up a consultation with an Auckland-based neurosurgeon, Arnold 'Ari' Bok. He was a South African by birth, and John assured me he had a very good reputation. Mr Bok turned out to be personable and very clear in his explanations. After considering everything, he told me he thought I had what is known as a meningioma of the cere-bellopontine angle—a growth involving the cells of the lining of a fluid-filled region in the rear portion of my head adjacent to my spinal cord. It was benign, which meant it wasn't going to spread to other areas of my body or invade other structures in my brain. It wasn't causing any trouble at this stage, but it was growing, and if left untreated it would definitely start to have bigger and more serious effects. Surgery looked like the only option that conventional medicine could offer me. But surgery came with risks, Mr Bok told me. Given where the tumour was and how big it was, there was around a 25 per cent chance that I would suffer deficit as a consequence of surgery.

'Deficit?' I asked.

'Negative effects,' he explained. 'In this region of the brain, loss of sensation to the left side of the face and fine motor skills. The worst case scenario would be where there was some damage to the spinal cord...'

You didn't have to be a brain surgeon to know what that meant. The consequences of damage to my spinal cord would range from paralysis to death. And what I was hearing while he was talking was that, at best, there was a 25 per cent chance that I would never fly again.

***

The confirmation of the diagnosis by Mr Bok was the beginning of a whole bunch of research. I hunted out the names and locations of anyone in the world who claimed to be able to treat this disease, and meanwhile, I obsessively researched the implications of each type of procedure. I quickly learned that the chances of a positive outcome were highly dependent upon whether the tumour was infra- or supratentorial. The tentorium is a kind of dividing layer between the greater part of the brain (above) and the cerebellum and spinal cord (below). Infratentorial tumours were much more treatable and recovery times were far less than supratentorial disease.

I would run each discovery past Mr Bok. After a while, Mr Bok became unavailable to talk to me on the phone, and his emails in reply to mine became more and more terse. Eventually, he wrote to my GP saying that he had explained everything to Mr Funnell to the best of his ability and the strong implication was that Mr Funnell should be encouraged to leave him alone.

This was a terrible time in my life. I had gone from being a productive commercial pilot and a member of a team that helped other people, even saved lives, to nothing. I faced Hobson's choice. If I did nothing, I would never fly again. If I submitted to surgery, there was a chance I would fly again, but there was also a chance that I would suffer dire consequences.

It was the first time in my life I felt powerless. Flying had been everything to me, and I couldn't imagine a future without it. The thought that I should perhaps jump off the town bridge and end it all crossed my mind, but that was the lowest moment. I quickly dismissed that as a loser's job, and rededicated myself to research.

I was fortunate to have lots of very good support, and not just from Trish. Dr Bryce Curran, an anaesthesiologist I knew quite well at Waikato Hospital, heard about my situation and drove down from Hamilton to have a chat to me. He told me he had known as soon as he heard what the loss of my pilot licence would have meant to me.

'I reckon you should read this,' he told me. He gave me a copy of Lance Armstrong's book, *It's Not About the Bike,* and he told me a little bit about the story. I'm not a great reader, but I was interested enough to have a look at Armstrong's book. And once I started reading, I was caught up in it. These days, he's better known for his systematic abuse of performance-enhancing drugs and his untruthfulness on the subject, but all of that has been allowed to obscure Armstrong's successful battle with testicular cancer that had spread to his lung and brain. Armstrong's major point was that it was his positive attitude that had saved his life. He was determined to beat the disease from day one, and he had no doubt that it was his determination that yielded a positive outcome.

Bugger it, I thought. If he can do it, so can I.

***

I don't recall who it was, but someone gave us the name of João de Deus (the name means 'John of God'), who practised faith healing in a little town named Abadiania, about an hour outside Brasilia. There were remarkable stories about him—people claimed to have gone in to see him in wheelchairs and left again walking, when conventional medicine had told them it could offer them no hope. Trish and I talked it over, and decided that it could do no harm to try.

We flew to Brazil, and made the journey to Abadiana. When we got there, we learned that João wasn't available, but that there were certain products we could buy—vials of water or packets of poppy seeds upon which he had bestowed a blessing, for example—which had been successful in others. I bought some stuff.

When I got home, my good friend Terry Burcher, who was a general surgeon in Hamilton, and who was on the Rescue Trust, asked me what I had been doing in Brazil. I was a bit hesitant to tell him, as he is a surgeon and would not, I thought, be inclined to think much of spiritual healing. But he shrugged, and told me that he had seen countless instances of people whose cases were, to all intents and purposes, identical—same age, gender, same disease and so on—who differed only in their attitudes, and that in every case their attitude seemed to influence the outcome. Those who had a positive attitude invariably did better than those who worried or expressed doubts about the treatment they were undergoing.

'I know you, John,' he said. 'You'll try anything, and good on you. It's the attitude that counts. You'll find what you need to get over this.'

Trish and I flew back to Brazil. This time, we met João, a fairly unprepossessing middle-aged fellow with thinning hair who spoke to us through an interpreter. He directed me to a room where I was to sit with my eyes closed and contemplate the cure of my condition. That was pretty much it. João is famous for performing what he calls 'visible' operations upon some patients up on a stage and in front of video cameras. But I was receiving what was known as 'invisible' treatment. It was faith—his and mine—that would do the trick.

***

In the course of the next few months, we consulted and interviewed neurosurgeons all around the world about possible treatments. Most of them were very interested in my case, and quite open about the fact that they would enjoy the challenge of operating on me. One procedure I kept hearing about was the so-called 'gamma knife', the use of intense, highly localised radiation to destroy tumour cells. I asked Ari Bok about it, and he conceded it might be an option. But it wasn't available in New Zealand, and he couldn't say who would be the best person to advise me on its suitability.

Then, one afternoon, he phoned to say that a Swede who had helped Lars Leksell develop the gamma knife technology was going to be speaking at a conference in Cairns. Bok had contacted him and he had said he was willing to talk to me.

So Trish and I hastily arranged tickets to Cairns. As we were just about heading out the door to go to the airport for our flights, a friend phoned and asked whether we were watching the television news.

I turned on the TV and saw the Twin Towers ablaze, and footage of an airliner disappearing into the glass of one of the buildings. I was too preoccupied to pay much attention. I thought it was a movie, and that these were remarkable special effects. It wasn't until later, at the airport, that we learned the terrible truth about what we had been watching.

We met the Swedish radiosurgeon in Cairns, and he looked at my scans.

'I helped to invent the gamma knife, and it is a wonderful tool, if I do say so myself,' he said. 'But do not let anyone try to operate upon you with it. It is a very precise tool, but with the tumour this close to the spine, the slightest mistake...'

He shrugged eloquently.

'What would you advise?' I asked, my heart sinking.

'Well, there are two people in the world whom I would recommend to operate upon such a tumour,' he said. 'It so happens that one of them, Professor Samii, will be at a conference in Sydney in two days' time that I am also attending.'

'Oh, if only we'd known,' I said. 'We could have made an appointment.'

'I am sure that if you are in Sydney when the conference is on, Professor Samii will see you. I'll tell him to expect you to contact him.'

I got straight onto Google. Professor Majid Samii was Iranian-born, I learned, and had trained in Germany. He now worked at the neurological institute attached to the hospital in Hanover, and thanks to him, this had become a Mecca for neurosurgical patients. Professor Samii had been many times honoured by the profession, and he was addressing the Sydney conference as the president of the International Society for Neurosurgery. Trish and I booked to stay at the Sheraton, which was hosting the conference.

It was almost possible to enjoy our next two days in Cairns, now that we were travelling hopefully.

***

Professor Samii phoned at 11.30pm on the evening he arrived for the conference. We had left a message for him, and the hours had dragged as Trish and I waited in our room, too scared to go out in case he phoned. He apologised for the hour, and agreed to meet us after breakfast.

After a fitful night's sleep, we woke early and spent the next couple of hours in a high state of anxiety, again afraid to stray far from the phone. We had neglected to ask what time he preferred to eat breakfast.

The call finally came, and he came to our room to meet us. He was a very dapper, well-spoken gentleman, with only the trace of a German accent to his English. He shook my hand warmly. He radiated complete self-assurance, and there was a trace of mischief in his eyes behind his glasses. I was inclined to like him straight away, and I reckon I knew from that moment that this was the man who was going to cure me.

He asked to see my MRI scans. I gave them to him and began to tell him what I understood about my condition. But he stopped me by holding up a hand.

'If you please, Mr Funnell, I will reach my own conclusions, and then we can compare what I know with what you know.'

He held the films up against the window one by one, nodding to himself.

'Well, Mr Funnell,' he said. 'It seems to me you have a meningioma of the cerebellopontine angle. This is a big operation, but it is completely treatable.'

'You can do it?' I asked.

'Yes, we can do this,' he replied.

'Do you keep statistics, Professor? What are the chances I would suffer deficit?'

'Without examining you thoroughly, I would say your chances are less than five percent.'

Trish and I looked at one another.

'Professor, your colleagues in New Zealand won't give me better odds than 25 per cent. Why are your numbers so good?'

He smiled.

'With all due respect to my colleagues in New Zealand, Mr Funnell, they would see one or two of these cases in their careers. Give or take an operation or two, I have performed this procedure twelve thousand times. We have pretty good systems at Hanover.'

'When could you do it, if we decide to go ahead?'

'We have spaces in three weeks or six weeks. I will give you my card. Please have a think about it. This is not a decision to take lightly, but if you decide you would like to proceed, email my secretary to make an appointment. And I assure you, Mr Funnell, we will be successful.'

Trish and I didn't need to think too long or deeply. We flew back to New Zealand, insulated within our own little bubble amid the chaos that had overtaken international aviation. And within a couple of days, we had our tickets to Germany and I had my appointment with the knife for six weeks' time.

The only slight hiccup was when the legal documents arrived. It looked as though we were being asked to provide a deposit of one hundred thousand euros. If that was the deposit, I gasped, what was the operation itself going to cost? But upon consulting Fritz, a German-speaking friend of mine, he was able to reassure me that something had got lost in translation. The figure was an estimate of the total cost of the operation, payable in advance. It might end up costing more, but it wasn't expected that any further sum would be significant.

After months of feeling helpless, I now had a definite plan, and it made all the difference. I immediately enrolled in a local gym and got put onto a special diet and exercise regime. In three weeks, I lost a couple of kilograms and lowered my blood pressure to levels it hadn't been near for decades. I was fit for the fight.

***

I had always kept pretty good health, and certainly had never had any experience of medical intervention on this scale. And while I had a pretty healthy respect for what medical science can do, having seen plenty of examples of it in action in emergency situations, I found it difficult to be blasé about the fact that a bunch of people would soon drill my head open and start fishing around amongst my brains.

Setting off for Singapore on the first leg of the flight to Hanover was a bit like setting out on the walk to the gallows.

We flew to Singapore and then onward to Frankfurt, before switching to a German domestic flight to Hanover, arriving there just before lunch on a lovely autumn day. There was accommodation for the families of patients at the hospital, so Trish was staying on-site. I had no complaints about my room, which I had to myself and which had a TV and an en suite.

Once we were settled in, we found our way down to the cafeteria to get some lunch. We were standing in line behind a middle-aged couple who were speaking English, albeit with a distinctive Scottish accent.

We introduced ourselves to them. It turned out Chris and Liz Wiper were both vets. They were there because he had an acoustic neuroma.

I asked why they'd chosen Hanover for their surgery.

'I did twelve months of research,' he answered. 'Professor Samii's reputation is unsurpassed. This is the place to be for my surgery.'

That was reassuring!

The following day, I underwent some tests with Professor Samii and his team. He greeted me warmly.

'Mr Funnell, your images are three months old. Best we get some up-to-date scans. Please present yourself to level one and have them done.'

'What,' I said, 'just like that?' It can take anywhere upwards of three weeks to get an MRI in New Zealand.

'Yes, just like that.'

I hardly had to wait before they invited me to lie down on the bench and fed me into the machine. It was all done within 20 minutes.

***

The staff shaved my head, and when I looked like something that had bolted from the Taumarunui shearing shed, they pasted a bunch of things the size and shape of Lifesaver lollies all over my naked scalp.

I had a chat with the anaesthetist, and she gave me a briefing and asked the usual ten thousand questions. I had a couple of my own.

'Doctor, will there be any pain?' I asked.

'No, zere vill be no pain,' she replied firmly. 'If you feel pain zen ve vill haff failed. You must tell us immediately, and ve vill manage it. Zis is most important.'

Soon I was alone with Trish.

From somewhere—from seeing head injury cases—I had got the idea that it would be a good sign if I woke up with a size 10 headache.

I had a horror of waking up to discover that something had been nudged during the operation and I had been left a vegetable.

'If it doesn't go well, promise me one thing,' I told her.

'What's that?'

'You'll take me over to Holland and get them to fix me,' I said.

'How will I know if something's wrong?' she asked.

'If I can't tell you not to, it's time to do it.'

A sad look appeared on her face. That was in the days when she thought that having me put down would be a bad idea.

***

Professor Samii came to see me in the evening. He looked reassuring, fresh and confident. He told me that everyone had a big day ahead of them tomorrow. There were only two operations scheduled. The Scottish vet was the curtain-raiser. I was the big game, and kick-off was 10.30.

Time dragged in the morning. They had said they would give me a couple of pre-med pills half an hour before the operation. It was closing in upon 10.30 when they arrived in a flurry and handed me the pills.

'I'm supposed to have them half an hour before,' I said.

'It won't matter,' they replied.

The bed thing I was on was an incredible piece of apparatus. I didn't have to do anything, and nor did they: with a whir of motors, it delivered me onto the gurney like a donut machine. I remember seeing the clock as they wheeled me into theatre, and it read just before 10.45am. I was surrounded by people with surgical masks on, trying to hide all the surgical apparatus from me. I recognised most of it.

They put a lure into my arm.

'If you are ready, ve vill now put you to sleep, Mr Funnell,' I recognised the anaesthetist's voice. I saw Dr Samii's eyes with their amused twinkle above his mask. I looked at the clock to see how long I could fight against the drug.

***

The light was dim and golden. I was floating. No, I seemed to be lying on something soft, and something incredibly soft and warm was enfolding me. Clouds perhaps.

My surroundings slowly came into focus. There was a young woman, blonde, blue-eyed, with impossibly clear and radiant skin, standing at the end of the bed. She was wearing white, and she seemed to be smiling at me.

Pain, I thought. There isn't any. I'm feeling nothing.

The soft warmth around me might have been a feather duvet. I moved a hand experimentally. I raised it to my head and felt it touch bandages. My head was swathed in them, as though I was wearing a turban.

Another sensation intruded on my flickering consciousness. I moved my hand in a southerly direction, past various wires taped to my chest, and found that I was catheterised.

But I looked back at the vision by my bedside. There didn't seem to be any room for doubt: the operation hadn't gone well. I had died on the table, and gone to heaven.

I wished they'd remembered to take the bloody catheter out.

***

I must have dozed. When I came around again, I realised I was in a bed, after all, and I was covered with a duvet. I was still catheterised, there was the drag of a line still in my arm, I had a nasal oxygen feed in, and cannulas draining through my nose, too.

Trish came around the end of the bed.

Shit, I thought. She's found a way to come to heaven with me.

She looked intensely relieved as I turned my head to look at her. There was still no pain, but I was inclined to think I was alive, after all.

I had asked her to see if she could come into recovery, and perhaps to take a photo of me as I lay there all intubated and wired up to the monitors. She produced the camera.

Poor Trish had a habit in those days of cutting people's heads off when she took photographs. So when I asked before the operation to try to get a snap of me in post-op, I lectured her on the basics of photographic composition. I didn't realise until a couple of years later that her camera had been dropped and the viewfinder no longer aligned with the lens. It wasn't her fault.

But presuming, as I did before that discovery, that she was simply a useless photographer, and imagining that I had made a more or less full recovery from the anaesthetic, I grabbed the camera off her and took two photographs of myself. She took a few more. Later on, she showed me the results. She managed to get some good ones. I got one of the bed and one of the ceiling.

Trish kept writing notes and showing them to me, because my hearing was impaired by the swelling after the operation and the dressing.

'Keep your voice down! You're waking everyone up! You're shouting!'

I drifted off to sleep again.

This time, when I woke up, it was around 6.30 the following morning. I was in the post-operative ward, with another beautiful young Fräulein there.

'How are you, Mr Funnell?' she asked. 'Is your throat all right?'

'Yes,' I replied.

'Would you like to brush your teeth?' she asked me, handing me a brand-new toothbrush, still in its crackling cellophane wrapper. 'Would you like cold water or warm water?'

Perhaps it was heaven, after all.

Professor Samii was there with a gaggle of junior colleagues and students. He smiled at me.

'The operation went very well, Mr Funnell.'

He did a few tests, rubbing my hands, holding up fingers to check my responses and so on.

'Very good,' he kept saying. 'Very good.'

He told the ward nurse to check me in a couple of hours. If my oxygen saturation and other indicators were still good, I could be returned to my own room.

***

By 10am, less than 24 hours after I went into theatre, I was back in my room. At about three in the afternoon, the door opened, and a strapping example of German manhood entered.

'Good afternoon, Herr Funnell. My name is Klaus. I vill be your physiotherapist, hmm? How are you feeling?'

'Good,' I replied.

'Right! Vere are your schlippers und dressing gown?'

I pointed them out.

'Do you want to borrow them?' I asked.

'Right!' He lowered the electronically adjustable bed. 'You vill now sit up und put on zer schlippers und zer dressing gown. Zis foot. Very gut. Und zis foot. Und now zer arms...'

'But I've just had a hole drilled in my head,' I began to protest.

'Ja. You may now valk about. Please, you vill now stand up!'

I meekly did as I was told.

'Gut. Now you vill valk to zer bathroom and vash zer face. Take care not to get zer bandages vet. You may vish to use zer toilet. You vill please to make sure you keep one hand on zer bed or on zer rail in zer bathroom. Ve do no vant any falls, hmm?'

I went to the toilet, and splashed some water on my face. When I got back into bed, he told me I could now choose whether I used a bottle or the bathroom. There was no reason why I couldn't move about.

'I vill be back in zer morning,' he said.

And he was as good as his word. At about 7.30 in the morning, there was Klaus.

'Right!' he barked. 'Stand up!'

As on the day before, he manoeuvred the bed to make it easy for me to dress and get up.

'Right! Follow me!'

We were seven floors up. He pointed to the stairs.

I began to point to the turban on my head, in case he'd overlooked the fact I'd lately undergone major surgery.

'Is okay! Zis is standard!' he yelled cheerfully.

We descended the stairs, flight by flight, with a 45-second pause that he timed with atomic precision on each floor. When we reached the ground, I presumed he would call a lift. Instead, he pointed to the stairs again.

'Very gut!' he pronounced, when we reached the top, and I had to agree. I felt amazingly like myself.

Klaus told me that I could use the stairs by myself if I wanted to, so long as I took a 45-second breather on each floor.

By day three, I felt so good that I suggested to Trish we go looking for an English newspaper. She reckoned it was about a 10-minute walk to the train station, and then a 20-minute train ride to the centre of Hanover.

I got dressed in jeans and T-shirt and asked the nurse if I could have a walk around the grounds. She thought that would be fine. So Trish and I went for a walk. By the time we reached the boundary of the beautiful, park-like grounds of the hospital, we were already a quarter of the way to the train station. Why not carry on?

We did it again the next day, sloping off to downtown Hanover to read the paper in a quiet café. No one seemed to notice or care, despite the turban I was still wearing. I felt the way I used to feel when I sneaked out of boarding school, way back when.

But on the third day, while we were in town, I began to feel the heaviness of exhaustion settling on me.

'We'd better go back,' I told Trish. 'I don't feel right.'

She suggested we get a taxi, but I thought I'd be okay taking the train. I fell asleep. She looked really worried when we got to our station. Again, she suggested we take a taxi, but I thought I'd make it.

Halfway back to the hospital, though, I knew I wouldn't. There was a little bar there, so I asked Trish to take me in there. I tried ordering a coffee, but the waitress didn't understand.

'Beer?' I asked in desperation.

'Ah, Bier,' she answered, and brought me a tall glass of crisp German lager, all dewy with condensation. It went down a treat, so I had another. After two beers, I felt a million dollars, and marched triumphantly back to hospital.

'Mr Funnell, ve haff been looking for you,' they said. 'Vere haff you been?'

As progressive as they seemed at Hanover Hospital, I didn't think they would approve of the two beers.

'Just for a walk in the grounds,' I said, hoping they couldn't smell my breath.

***

The turn I'd had and which I'd fixed with beer didn't recur, and I decided that because I had been on a special, low-sugar diet, I had simply let my blood-sugar run too low. I felt fine after that, and was discharged two weeks later. I wasn't supposed to board an aircraft for another two weeks after that, and they gave me the option of staying on at the hospital or shifting over the road to a hotel. I chose the hotel, and we booked in there. As luck would have it, there were issues with our own hotel at home, and I gave Trish the green light to head home and deal with them.

Along with the vet, whose own recovery had been slower than mine, I attended a couple of lectures that happened at the hospital: Hanover is a teaching hospital attached to the university, and a couple of mini-conferences took place in that two-week period. Between him and me, with our respective work experiences and the research we had done into our own conditions, we managed to get a lot out of these highly technical presentations. Everyone else sitting in there had multiple medical degrees. Our turbans were our qualifications.

The rest of the time, I visited museums and did some sightseeing. There were times when I wished Trish was still there, or that I'd got her to send Hamish or Mark over: my hearing loss made it difficult at times. Similarly, when I got a train across to Switzerland to have a bit of a look around, the effort of dragging my bag around took its toll.

Even so, just four weeks after I went under the knife, I did a kind of exit interview with Professor Samii. He expressed his general satisfaction with my recovery.

'Professor, will I ever drive again?' I asked him.

'Drive?' he said, startled. 'Mr Funnell, you can fly your helicopter. I would ride with you with no concerns at all.'

I couldn't believe what I was hearing.

'Could I get you to write a letter?'

'Of course! See my secretary on the way out. She will provide a letter, and copies of the academic material I have published on our procedures and their impact upon brain function.'

***

I flew home, and armed with this material from Professor Samii, I immediately applied to the Civil Aviation director, Captain John Jones, for information on how to seek the reinstatement of my pilot's medical.

'You'll need to go and see Bill Wallace in Auckland. He'll do an assessment for you,' Captain Jones said, and the new CAA principal medical officer Dougal Watson confirmed it.

So I made an appointment with Mr Wallace, an American neurologist working in Auckland.

He sat up when I mentioned Professor Samii's name.

'He's the top man in the world in neurology,' he said. 'How did a Kiwi get to see someone like him?'

'Well,' I shrugged, 'it was easier getting an appointment with him than with Ari Bok at Mercy here in Auckland.'

I talked him through the procedure that Hanover had performed, especially the nerve monitoring technique they had used to avoid inflicting damage on nerves adjacent to the operating site. I showed him the fat folder of academic literature I had on the subject.

'How did you get hold of these?' Wallace asked.

'I asked nicely.'

'I'd love to...'

'They're yours.'

Almost five months to the day after my surgery, in May 2002, I received a letter and my pilot licence. Mr Wallace had written to the CAA to recommend full re-instatement. There was a cartoon of a brain with a dotted line and 'cut here' emblazoned across it. But the real McCoy with my photograph arrived a couple of days later.

I was back!

Landing at Banda Aceh to deliver aid.

CHAPTER TWELVE

BANDA AID

A day or so after Boxing Day 2004, I got a call from Dave Kershaw, operations manager for a company named Hevilift, based in Papua New Guinea. They were under contract to Oxfam, who were desperately trying to get aid to those in need after the earthquake and tsunami that had devastated the Indonesian province of Aceh. They could only spare one Mil-8. They were aware that we had Mil-8s, and wondered if we could spare another to assist in the aid effort.

Before I left, Trish gave me a couple of 25-kilogram bags of lollipops, which she thought might go down well with children in the affected areas. And she told me that if I happened across an orphan or two, she wouldn't mind at all if I brought them back.

We quickly assembled flight crews, fitted long-range tanks to the machines and set off on the 4600-nautical mile (8500 kilometres) journey, flying from dawn to dark. From Taupo, we flew to Kerikeri, to Norfolk Island, then to Lord Howe Island, then to Roma in North Queensland, then Darwin, across to the Indonesian island of Java, along to Sumatra.

We arrived at Medan, the closest unaffected city, and spent the night there. Then we flew on to Banda Aceh, where we were to be based, arriving four and a half days—47 hours' flying—after receiving the call. We were apparently only the second civilian helicopter on the scene.

It was odd, but everywhere we set down in Indonesia, people were completely incurious about our mission, and didn't seem too bothered by what had happened in their western provinces. We began to realise that to add to everyone's woes, there were political and ethnic divisions on Sumatra to contend with, quite apart from the aftermath of the disaster itself.

At Banda Aceh, the airport had become a tent city. The Australian Army had set up across the way from us. You couldn't buy anything—everything was coming in by 737—but we had brought our own supplies of bottled water, freeze-dried food donated to us by the very public-spirited company Backcountry Foods of Invercargill, and a gas cooker. We met with the Oxfam people. You could tell from their faces that what they were dealing with was beyond words. They told us that their organisation's principal goals in the relief effort were providing water and sanitation. Nothing was left of the towns and villages in the affected area, they told us. It had been a poor area to begin with, but now everyone was starting from scratch. We were assigned a driver and a translator. The translator was a slightly built man named Marwardy, who had studied to university level and who spoke English with a startlingly correct BBC accent. There wouldn't be much driving involved in the affected area, as the bridges had mostly been destroyed, but our driver would be invaluable in helping us survive Indonesian traffic in the areas where life was going on. We got an early taste of this as he drove us to the hotel where we commandeered rooms. Indonesians seem to take a Darwinian approach to the business of driving. If you get the nose of your vehicle in front, you rule the road. There's an expectation that everything else is sorted out by leaning on the horn. It seems to work: we didn't see any crashes, but we were glad to make it alive.

We began drawing up our operational plan.

Before we flew our first mission, we made a list of the various risks we might face. There were the run-of-the-mill aviation issues to consider, but some were more acute due to the situation. The risk of fuel contamination was heightened, as was the risk of rotor blade strike on personnel unfamiliar with helicopters running about the place. There was likely to be lots of loose material in our landing zones, and we had to take care not to draw debris into the engine intakes.

'Birds,' someone said. 'If there are lots of bodies around, there'll be lots of birds.'

That was a good point.

'Bear in mind,' someone from Oxfam said, 'if you're flying over this area, you'll be flying over areas controlled by the political opposition. They unfortunately associate the relief effort with the government. It might pay to keep out of range of small arms fire.'

'Jesus,' we muttered, and added a directive to keep above 3000 feet when overflying the areas he indicated to the list.

Then there were the other risks. It was bloody hot and humid: we would have to make sure we proactively avoided dehydration. The contamination of all food and water should be presumed, unless the water was in a bottle or the food was in a packet or needed to be peeled. The risk of contracting infection in any wounds sustained was high. It was a tragic fact, the Oxfam people told us, that a large number of the people who had survived the disaster were going to die from malnutrition or diseases such as dysentery or infections from wounds.

We were starting to get a sense of what it would be like.

'One more thing,' the Oxfam guy said. 'For Christ's sake, stay away from the local women. They'll probably be keen, but if you're caught messing around with them, their brothers will give you a hiding and make you marry them.'

I don't know what nonsense he'd heard about helicopter pilots to think it was necessary to pass on this bit of advice.

***

We loaded our first payload—water pumps and purifying gear, mosquito nets, bedding, medical supplies and a 4WD vehicle—boarded our crew and a couple of Oxfam people and lifted off.

For a few kilometres south of Banda Aceh, the countryside unrolling beneath us was much like everything we had seen along the coasts of Java and Sumatra: palm plantations and crops interspersed with villages and muddy roads. But then, fully 8 kilometres inland from the shoreline, it was as though someone had taken a giant scouring pad and made an effort to erase all sign of human habitation. Crops and plantations had disappeared, leaving only bare earth. Villages had gone. Only concrete foundations of buildings remained, and the rest was just a jumble of building materials with other human artefacts—cars, trucks, motorbikes, the corpses of farm animals—tangled up amongst the drifts of wrecked vegetation.

As we began to descend, we could see raggedly dressed people running amongst the debris, desperate to get to our landing site.

'We've got to set up a perimeter when we're on the ground,' I briefed the crew. 'We need to keep people back from the rotors.'

So as soon as we touched down, everyone in the helicopter jumped out and did their best to keep people back while I stopped the rotors as quickly as I could. The landscape had been covered with fine silt which had dried, and it flew about us in choking clouds. It was hot. The heat blasted in as soon as the doors were open, and as soon as you got out, you were assailed by the stench of decay, both animal and vegetable.

Upwards of three hundred people might be thronging the machine at most of the sites where we landed. They were very respectful, and very keen to help us unload. We distributed water, but there was still a look of pinched and desperate pleading on every face. Marwardy told me that the people expressed their gratitude for the water, but said they needed food.

I had stuffed the deep pockets of my overalls with lollipops, and as inconspicuously as I could, I sidled up to children who were standing at the edge of the crowd and while pretending to shake hands with them, I would slip them a lollipop. It was a joy to see their eyes widen when they saw what I had given them. It didn't immediately occur to me that it was the first sustenance of any description they had had for days, perhaps even since the ocean came ashore 14 days earlier.

At lunchtime, back at base, we cooked up a sumptuous feast of two-minute noodles. We were sitting eating them when I noticed Marwardy and the driver sitting disconsolately at a distance, trying not to watch.

'You guys got any lunch?' I called out to Marwardy.

He shook his head.

'We have not eaten for a few days,' he said.

'What? Nothing?'

He shook his head again.

Never were instant noodles so deeply appreciated.

***

We flew two or three missions a day, carrying loads of up to 4 tonnes at a time. In order to refuel, we had to use cash, and only American dollars would do. Each fill would cost $1000, so I had wads of cash stashed all over my person and the aircraft.

You occasionally hear criticism of the big aid organisations and the costly and unwieldy bureaucracy that clips the ticket between well-meaning donors in the West and the needy in disaster areas. But it was hard to feel anything but the deepest admiration for the people I worked with on the ground. Oxfam in particular were very focused and practical.

But the most admirable person I met there, and by far the most efficient distributor of aid, was a Kiwi guy, a doctor from Gisborne who turned up with his private two-seater helicopter and a supply of past-use-by-date medicines he had cadged from doctors and hospitals back in New Zealand and was personally distributing. There wasn't much ticket-clipping going on in his operation!

Every now and then, a dump truck would drive past loaded to the top with bodies destined for one of the many deep pits that had been bulldozed to serve as mass graves. There was no effort to identify people: six days in the tropical heat meant no identification was possible. They were simply dumped unceremoniously in the ground and covered as quickly as possible. And for every body recovered and thus disposed of, you knew there were two or three still buried in silt or in the wreckage of buildings.

One day, Marwardy took us to his former family home, now just a jumble of broken concrete.

Marwardy was a surveyor by trade, and he told us that on the day of the tsunami, he had been away working, further east. His sister, his brother-in-law and their young children had ridden into town after the massive earthquake, the epicentre of which was just 300 kilometres east of Banda Aceh. Along with Marwardy, they had escaped both the devastation caused by the earthquake and the deadly tsunami that followed. But Marwardy's mother had an injured leg, and wouldn't have been able to run. His grandparents were too old, and his father wouldn't have left them.

'My mother, father, my grandparents and my niece are buried in there,' he said without emotion.

His two brothers had actually been caught by the inrushing water and washed onto the second floor of another building. They had survived.

I asked Marwardy how his sister was feeding her children.

'She must use rice water to feed her baby. Formula milk is too expensive,' he said. He gestured at his shorts, T-shirt and sandals. 'They are like me, Mr John. This is all I have now. Everything else is gone.'

I reached into my pocket and peeled US$100 from the roll of notes I had there and handed it over. He took the note and just stared at it, his eyes wide with disbelief.

'You know what you've just done?' one of my crew asked me after he'd gone. 'You've just turned him into the local millionaire.'

He was probably right.

***

Towards the end of our time there, we were working at a village that we had been using as a waypoint when Marwardy tugged on my sleeve.

'Mr John. The village chief would like to talk to you.'

He pointed to a ragged-looking man who was smiling at me a short distance away with three giggling girls a little behind him.

'Does he speak English?' I asked Marwardy.

'No.'

'Then you'll have to come with me. What does he want?'

'He wants to give you those girls,' Marwardy said.

The man nodded and grinned at me. The girls, who were probably somewhere between 13 and 15, giggled like mad.

'What do you mean?'

'He is very grateful, Mr John, and he wants to give you these three girls.'

I wondered what nonsense the village chief had heard about helicopter pilots, to make such an offer. And I wondered what Trish would say if I turned up at home with three pubescent Sumatran girls and said: 'You know how you said I could bring back some orphans...?'

'Tell him,' I said carefully, 'that I am very grateful but I cannot possibly accept.'

'Mr John, you will cause great offence if you refuse,' Marwardy pleaded.

'Believe me, it will be nothing compared with the offence I will cause at home if I do accept,' I told him.

Through Marwardy, the man made repeated attempts to get me to reconsider. But after giving the girls a handful of lollies each, I excused myself and fled.

***

When finally we left, I gave Marwardy a wad of US cash and told him to look after himself and his sister's family. Since then, I have also brought him out to New Zealand, where he has attended a language school to advance his English. When I've done presentations to service clubs and similar groups, I've dragged him along, and although he hasn't enjoyed the experience, I've got him to submit to questioning about the slow but steady progress his family and his people have made since Boxing Day 2004. It's unimaginable what they have gone through, but quite humbling to see the resilience of human beings, too. People have tended to give generously.

I am haunted by Aceh. I have never seen anything like it, and I hope I never see anything like it again. It felt good to have been part of the effort to make a difference, even if the difference we made looked and felt like trying to put out a forest fire by pissing on it.

***

It was typical of my son Mark that when he decided he was going to start jumping out of aircraft he didn't tell anyone about it. Knowing how Trish and I would react, he just took himself off and did it. Both boys had followed me into aviation, and went on to get their commercial fixed-wing and helicopter licences. Mark started flying the skydive plane for Taupo Tandem Skydiving.

One day, our chief engineer Tom McCready was at the drop zone talking to him. He leaned on the aileron of the plane and it swivelled and moved inwards and stopped with a big 'clunk'. It shouldn't have been able to do that.

'What the hell are you doing flying this thing?' Tom said. 'It's badly maintained.'

Mark—typical Mark—shrugged.

'It's a flying job. And, anyway, I've got a parachute. If anything goes wrong with the plane I can just jump out.'

'Fat chance of getting out of a plane if the aileron comes off,' Tom pointed out. 'You'll be pinned to the floor by the centrifugal forces. I'm going to tell your old man about this.'

Tom did tell me, and he told me he hadn't been able to make much impression on Mark. That was typical.

I mentioned my disquiet about the skydiving outfit to my PA at the time, Helga Sonier. She knew Vaughan Brady, the operator.

'He wants to sell. Maybe you should buy it.'

This was an interesting idea. Mark had left the company and was off flying for Eagle Air by then. But I talked to Hamish, who was at university at the time studying toward a marketing degree. He was quite excited at the idea of perhaps taking a business and building it up. We agreed that we'd need to check it out, and that probably meant learning about the whole skydiving thing.

So we enrolled in a skydiving course with Taupo Tandem. The teaching department was Tim Russell, a very experienced jumpmaster who was getting pretty long in the tooth. We needed to do 25 jumps to get our licences, but it took forever to get them done. Tim tended not to like getting up too early in the morning. Quite often, we would wake up to glorious conditions—cool, calm, still—only for it to cloud over or start blowing by the time Tim finally came on deck. He'd look up into the sky and shake his head. 'Nah. Too windy/cloudy/cold...'

When we did go up, I found the whole thing terrifying. Partly this was because jumping out of a perfectly satisfactory aircraft at altitude goes against a pilot's deeply ingrained instincts, never mind your basic human survival instinct. Partly it was because Tim's method of telling you it was your turn to go was to smack the top of your helmet—it always made my teeth rattle in their sockets—and yell: 'Yeehaa!' That did nothing for my nerves. And partly it was because all the gear was so dilapidated. In today's environment, you probably would have made a bonfire with it a thousand jumps previously.

Trainee skydivers use what is called a static line system, which means that instead of opening the chute yourself en route to the ground, a line fixed to the plane activates a mechanism that deploys the chute soon after you jump out. You have a reserve chute that you can activate manually, in what is supposed to be the unlikely event of the failure of your main chute. By the time I had done 20 jumps, I had already had three reserve canopy rides. The gear was in a pretty sad state.

I got to know quite a few of the guys. There were a couple of Dutchmen, pretty typical of the sort of person who is attracted to this activity: young, fit, laid back and totally hooked on adrenaline. They did tandem skydiving during the Netherlands summer and then flew south for the winter, doing summers in New Zealand. I was chatting to one of them as he packed his canopy. There was a hole in it I could have stuck both fists through.

'René, there's a bloody great hole in your canopy!'

'Doesn't matter,' he shrugged. 'I'm keeping an eye on it.'

'We wouldn't fly an aeroplane with a hole in the wing...'

'Ah, it'll be all right.'

Most of the staff were aware that their operation was very run down. A couple of them dropped by and looked at our helicopter operation, and they were impressed by our maintenance standards.

'You should buy the skydiving company,' they told me. 'It could use a bit of TLC.'

Helga and I talked to Vaughan. It was clear he was looking for an exit. The company had lost a tandem skydiver and his client in 1998—they had jumped and had a malfunction and the reserve chute had tangled with the main canopy. The general opinion was that this episode had knocked the stuffing out of Vaughan.

He ran the entire operation out of a bum bag which he wore in front. I asked him what he thought I would need to do to expand the company.

'Nothing,' he said. 'Everything's sweet.'

I got Helga to open negotiations with him, and we'd soon agreed a price. As soon as we took over, I went to an Auckland firm that manufactures tandem canopies, so that at least we started out with the boys under good canopies. There's nothing for staff morale quite like making sure there are no holes in the flimsy bit of fabric that's supposed to stop them plummeting to their deaths from 15,000 feet.

Hamish got his commercial pilot licence and came to work with us, too, so I had both boys working for the company. The company had two aircraft, a Cessna 206 and a Beaver (a blast from the past for me). They weren't ideally suited to the work they were being asked to do, as piston-engined aircraft don't really like going to altitude. We replaced the Cessna with a Nomad and had the Beaver rebuilt.

We had quite a bit of work to do to change attitudes amongst some of the old hands who came with the operation. Because everyone was paid per jump, after four or five jumps some of them felt they'd earned enough for the day and would make excuses as to why no more jumps would be possible. But most who felt that way soon moved on, and were replaced by a younger, keener brigade. It also became apparent to me that some of the people we were employing were regular partakers of illicit substances. This worried me a fair bit. I had a chat to Hugh Barclay, our in-house solicitor, about it. He shared my concerns, and went off to do a bit of research into our level of exposure if there were to be an accident in which drugs were found to have been a factor. Unsurprisingly, he learned that it would probably mean the end of the company. So between him and me, we drafted a policy with regard to drugs, and called a meeting of all of our staff at home one night. We gave each of them a copy of the draft policy and explained the rationale for it: most of them had financial commitments of one sort or another. If we stuffed up with drugs involved, then that was everyone's livelihood gone. I wasn't prepared to accept that risk. They had a week to have a look at the policy and get any comments to us. Then it would become a company requirement.

Of course, we received no submissions. Who knows how many of them even read it. But this didn't mean there was no substance abuse going on. When we reconvened the meeting and told them that the policy had now been adopted, we announced that we were going to start drugs testing. I said that if anyone was a drugs user, they had two choices. The first choice was to keep doing it, and try your luck.

'You will be caught,' I said. 'You'll have two strikes. If it happens again, strike two, you're down the road. On the other hand, if you come and see us in the next 24 hours, we'll undertake to help you deal with it.'

Once they decided that we were deadly serious, one of them came and handed in his resignation. He just couldn't see himself doing what he did without using drugs. A few others tried it on. We caught one guy in the testing regime, and when we suspended him, he asked to come and talk to us about it. He came to the office on a Saturday morning and he was clearly pretty wild about it. Hugh was there, sitting at a desk writing. I explained that the rules under which we'd hired him were that if he tested positive for cannabis, he had to be suspended. If he straightened himself out, he could come back to work. But if he tested positive again, he was out. Without further ado, the guy took a swing at me. I saw it coming, ducked and then slammed the door in his face.

Hugh just kept on writing.

'What are you doing?' I asked.

'Just putting the finishing touches on a letter to him,' he said. 'I'm telling him that we take the punch he threw to be an offer of resignation, and that we accept.'

There was a lot of resistance to the policy at first—several tried to persuade me that cannabis doesn't impair your ability to do their job at all, and even offered to give me some to prove it. I wasn't having a bar of it, to coin a phrase. But after 12 months or so, the policy had bedded down and everyone had embraced it. In 2012, the prevalence of drug-taking in the adventure tourism industry was thrown into the spotlight when a hot-air balloon collided with a set of power lines in the Wairarapa and all aboard died in the subsequent fire. The operator was found to have been using cannabis before the flight. I was pleased that we had been proactive in this regard.

In freefall at Taupo Tandem.

Scott Donaldson in his kayak, 300 nautical miles out from New Zealand.

CHAPTER THIRTEEN

ON THE ROCKS

In the early hours of the morning of 5 October 2011, a large, Greek-owned container vessel, the MV *Rena,* inexplicably drove into Astrolabe Reef, a large and well-charted maritime hazard around 15 nautical miles from her destination, Tauranga. Conditions weren't particularly bad, but the Rescue Coordination Centre thought it would be best to have helicopters standing by to evacuate the crew if the ship started to sink.

We were there within the hour, and we spent the morning twiddling our thumbs. It turned out the vessel was stuck fast, and while she was leaning on a pretty rakish angle to starboard, she didn't seem in any imminent danger. The Korean crew were happy to remain aboard while the chances of refloating the vessel were assessed.

It was a different story a week later. The weather deteriorated, the ship rolled to assume a list to port and a big crack appeared in her topsides forward of the superstructure. About 80 containers broke their restraints and toppled off the deck into the sea. The navy's support ship *Endeavour* was on station near the wreck, and the crew were evacuated to this vessel. It was now looking highly unlikely that they would be able to refloat the ship, so the focus was squarely upon mitigating any environmental impact and salvaging as much of the cargo—1368 containers of assorted goods, including some that were classed as hazardous—before she broke up.

The company that successfully tendered for the job of removing the bunker oil and the containers was Svitzer, a Dutch company with a presence in Australia. We had worked with them before: in 2004, we sent one of the Mil-8s up to Futuna in the Wallis and Futuna group, about halfway between Fiji and Samoa, where another container ship, the *Southern Moana,* had run aground. The chopper was used to lift empty containers from the deck of the ship to lighten her, with a view to pulling her off, which was eventually accomplished.

Svitzer contacted me—we had returned home—and asked whether we still had a winch in the helicopter. They wanted us to act as shuttle bus for their salvage crew and equipment, because they had determined getting to and fro by boat was both time-consuming and dicey, especially when the weather was bad.

'When do you want the machine?' I asked.

'Well, now,' they replied.

We were there the next morning, and we started work just about immediately.

A typical day would start at 6am. We'd have a 20-minute briefing, and then we'd start flying at seven. We'd take groups of four out and we'd winch them one by one down to the deck. Just about every surface on the vessel was stacked high with containers, so there wasn't much room to land people or gear.

At the end of the day, we'd start flying at around 6pm, and we'd keep going until dark. The last man off was always Dave Skola. He was a hard worker, that man. Once the little oil tanker *Awanuia* arrived to receive the bunker oil that was pumped off the *Rena,* a skeleton crew of three was kept on the ship overnight to carry on pumping. We would take them off in the morning when we landed the daytime crews, and we'd fly takeaway food out to them at night as well.

It was plain that everything would be a damn sight easier, cheaper and more efficient if we could actually land on the ship rather than having to hover and take people and gear on and off with the winch. I suggested to Dave that we construct a helipad for this purpose. He designed one to bolt onto the top of a stack of six containers on the forward section; it was prefabricated ashore, and the large Bell helicopter that was flown out for the job from Australia took it out and set it in position where it was to be affixed. Once this was commissioned, everything was a lot faster. One day, when the sea was calm, I was able to leave the helicopter on the pad and get a guided tour of the vessel. It was an eye-opener. It was very interesting seeing the bridge, and the conditions in which the crew of a huge ship like this live and work. It was also an insight into how the salvors were working, with everything on a 22-degree lean.

Once the bunker oil was off, the focus changed to shifting containers. This was going to be dangerous work. Svitzer asked me whether I could recommend a risk consultant. It so happened I could. I put him on to Geraint Bermingham of Navigatus Consulting, and he arrived and began putting together a Safety Management Scheme for the container salvage. A list of hazards was drawn up, and each member of the salvage crew was issued with a passport.

'You're their only way on or off the ship,' Geraint told me. 'It's your job to check that they have the necessary checks in their passports.'

He jerked his thumb toward the roomful of tall, muscular Dutchmen in the meeting room.

'Well, I'll try,' I gulped.

As it turned out, they were a pretty well-disciplined lot, and quite used to working to plans of this sort. Theirs is a dangerous occupation, and often it is the systems that are in place that guarantee everyone comes home after their day's work.

Pat Wynne came out and was stationed on the *Smit Borneo,* a big barge that was handling most of the containers that were lifted from the ship. We were on standby in case there were any emergencies: I had got Hamish to fly another machine over so that we could offer continuity of service and to reduce the chances that another operator might duck in and take the work out from under us.

***

In January, everyone watched with a bit of trepidation as a big storm rolled in. Progress in shifting the containers had been quite good, and most of the deck cargo had been removed from the stern section. We had shifted the helipad to the prow, because most of the work was now being done on the forward section. But work had to come to a stop as big seas started battering the vessel. It was clear the stern section was coming adrift. On one of my flights, I had been able to see quite clearly that the hull was full of water. But although they had huge pumping capacity out there, they decided for some reason not to use it.

On 10 January 2012, we were having a meeting aboard the *Smit Borneo,* which was at anchor in Tauranga Harbour, when we heard that the life rafts on the superstructure had inflated.

'What does that mean?' I asked.

'It means she's sinking,' they replied.

Hamish and I both flew out there as efforts were made to tow the stern half of the ship. The big tug *Go Canopus* had a line on a section that had broken free and was trying to manoeuvre it into deep water. I was acting as a comms relay, and I was conscious that it was a potentially very dangerous position to be in. I would relay communications with carefully worded prefaces such as: 'This is an instruction for the master of the *Go Canopus.* The salvage master would like you to...'

As the situation deteriorated, the communications started getting pretty tense. I was stuck in the middle.

I actually sat my machine down on the helipad on the prow, and I had a grandstand view of the rear half of the vessel slipping underwater. It was a strange, impressive and slightly melancholy sight.

***

Once as many of the containers as possible had been taken off, Svitzer's job came to an end. There was a call for tenders to carry out the next phase of the operation, which was the removal of the massive wreck itself—or at least, that section of it that remained on top of the reef. The stern section had slipped further down the undersea slope and was well submerged, with about 200 containers trapped aboard. The successful tenderer was an American outfit called Resolve Salvage and Fire, and it was our good fortune that they came to us for the provision of helicopter services, too. They had also contracted the Bell 214 belonging to McDermott Helicopters of Australia, which they used to lift each section of the steel hull as it was cut away. It was our job to ferry the men who did the actual cutting and the gas bottles they used for their big gas lances out to what remained of the ship. They chopped the thing up, bit by bit, and they even had divers cutting it up below the sea's surface, too. They carefully cut around our helipad until it was getting in the way, whereupon they cut that up and took it away, too. After that, we were back to winching people up and down from whatever suitable bits of naval architecture we could find. When finally in July 2014 the salvage was over, it had been a steady source of income for us for two years. It was pretty good for the company, all told.

***

Around 2009, the Philips Search and Rescue Trust told me that there would be a significant tax advantage available to them if they owned the assets—the helicopter and rescue gear. Looking at their numbers, it was hard to argue. It made perfect sense to sell the assets to the trust; I would simply manage them on their behalf. We did the deal in 2010.

It all made perfect sense, but it wasn't ideal from my perspective. Since I was no longer the owner, I was no longer in a position to call the shots, and the way they wanted things done was not always the way I would have done them, left to myself.

My management contract ran for two years. As it came up to the end, almost 25 years to the day since the Philips Search and Rescue Trust had first been established, I made the hard decision that I would walk away. By this time, I wasn't really doing any of the flying at all, although from time to time I would still be asked to assist where technically difficult jobs arose, especially when they involved long-range flying.

In 2010, for example, I heard of the plight of Shaun Quincey, who was trying to emulate his father's feat of rowing solo across the Tasman Sea and who had struck trouble short of the New Zealand coast. He'd capsized and had lost a fair bit of his food, and more crucially, his desalination plant had malfunctioned. I approached his supporters and told them that I could probably carry out a supply drop, if necessary.

Quincey had got himself within 300 nautical miles (555 kilometres) of the coast, so he was within the range of my Piper Comanche. Hamish and I sat down and worked out the logistics of the whole thing. You couldn't just tip containers out of the aircraft and let them land in the sea. They'd need to be let down gently. But deploying parachutes from an aircraft is a dangerous business. If they wrap themselves around the tail-plane, you're in trouble. Hamish and a skydive rigger who worked for us, Brad Rock, devised a very cunning delay apparatus using a static line. The parachute—actually just the small chute or drogue that operates a tandem canopy—would be packed in a little container attached to the side of the drum which held the supplies. After it had dropped clear of the plane, the static line would pull out a pin that would release the chute.

We ran a few tests at Taupo airport, and decided the whole thing worked pretty well. Initially, we were going to fill the empty space in the container with polystyrene, but we realised that would be an environmental disaster. It would be a better idea to put useful stuff in there instead. I wondered what you would most want in the world if you had been away from civilisation for ages and were starving.

Roast chicken.

So we crammed two or three roast chooks in there, too, along with coffee, two copies of *The New Zealand Herald* and a toothbrush to go with 39 litres of water, which ought to last him 20 days.

We flew to New Plymouth, refuelled and then struck out over the sea. After two and a half hours of flying over the featureless Tasman, we found the little orange vessel. Sea conditions were pretty good. Mark had the controls, and I did the release. The parachute delay system worked a treat, and the container landed 20 metres away from him. We watched Shaun pull in the line that was attached to the container and circled until he had it all aboard his little boat. Then we headed back to the mainland. It was a nine-hour round flight.

Shaun called his girlfriend that evening in tears on his satellite phone.

The roast chicken was the most wonderful thing he had ever eaten, he reckoned, but she was horrified. Given how long Shaun had been rowing on short rations of a special diet—mostly porridge—the chicken would be quite a shock to his system, never mind the risk of food poisoning. We hadn't expected him to scoff the lot in one sitting, so we were a little concerned for a while, wondering if we might have sabotaged his mission with our good intentions, and perhaps even put his life in danger. But he didn't suffer any ill effects.

We were obliged to make another drop to him a few days later. On both flights, as we spoke on the marine radio while we circled overhead, it was plain he was desperately glad of the company. I was impressed with his positivity. I didn't have any real doubt that he would make it all the way.

Over the next week, though, instead of making progress towards New Plymouth, he was swept steadily north. Finally, in mid-March, he looked as though he would make landfall on Ninety Mile Beach in the Far North. We headed up to be there when he came ashore. James McOnie of RadioLIVE's Weekend Sport show was one member of the scrum of media that was amassing to meet him, and he had the bright idea of giving Shaun a bacon sandwich as he came ashore. We undertook to organise the sandwich, and so in the pre-dawn hours of 14 March we were scrabbling about our motel room frying bacon and buttering bread.

Ninety Mile Beach was like Auckland's southern motorway for a while there, as friends, family, supporters, well-wishers and media tore up the sand in 4WD vehicles to get to the spot where Shaun was expected to make landfall. Once he had been spotted, a mate of his on a jet ski went out through the breakers to escort him in. Shaun's boat, the *Tasman Trespasser II,* was designed to handle surf, but there was a substantial swell running. Rather than be rolled around in his boat, he abandoned ship and swam the last 400 metres to shore, as the surf lifesavers who were in attendance kept an eye on him from their rubber duckies.

Shaun strode ashore looking a million bucks, 54 days after setting out from Australia, and promptly sank his teeth into the bacon sandwich.

***

Around 2014, I got a phone call from a lady by the name of Sarah Donaldson.

'My husband is trying to kayak his way across the Tasman,' she said. 'He's run into difficulties, and he's getting low on supplies. We've got all this food to get to him. I heard that you can do supply drops out to sea.'

'Well, we can,' I said. 'How far out is he?'

'About three hundred kilometres,' she replied. 'Can you do that?'

That was a little bit over 150 nautical miles, or just over an hour's flying in my Piper Comanche. Adding the distance from Taupo to the coast—another 45 minutes—that was the guts of a three-and-a-half-hour return flight, without all the manoeuvring required to make the drop.

'In theory, yes, we can do it,' I replied.

I had heard of Scott Donaldson and his attempt to kayak the Tasman. I had heard that he'd had problems with the electrical systems on his sophisticated kayak and had been forced to put in to Lord Howe Island. It had taken a while to fix them, but he had got under way again. That had been weeks ago.

Sarah told me that if I looked on Scott's website, there was a tracker reporting on him and I could see his exact location. I looked him up and plugged the latitude and longitude into Google Earth.

'Er, Sarah,' I said. 'You were right about the three hundred. But it's nautical miles, not kilometres.'

'Is there much difference?' she asked.

'Oh, yes,' I replied. 'Three hundred nautical miles is double the distance offshore. That's going to be more than ten hours' return flight.'

'Can you go that far?'

'Well, it's possible...'

'Yay!' she said. 'So you can do it!'

'Hang on. Even at mate's rates, that's going to be bloody expensive.'

'How much are we talking?'

'Taking into account all the gear we'll have to use and never get back, about $10,000.'

'Cripes! That's the other problem, you see. We don't have any money left.'

'Okay,' I said. 'So what are you going to do? Just leave hubby out there?'

'I'd like to get a boat out to him.'

'It's three hundred nautical miles, Sarah. You're not going to get much of a boat for $10,000, and certainly nothing that can go that far out.'

'Oh,' she said.

'Leave it with me,' I told her. 'I'll do some thinking.'

I phoned Ruth Wynn-Williams, a reporter I knew at Television New Zealand, to ask if we could do some sort of deal with them. They were interested, but only if they could get good footage out of it. It took three days of negotiation, but in the end what we worked out was that if they put the pictures we took to air, then they would pay—not the full amount, but enough to pay for the fuel we'd use.

I called Sarah to tell her the good news, and we started planning the operation.

A box arrived from Backcountry Foods in Invercargill—it turned out that Donaldson's venture was yet another cause they were supporting, without asking for any publicity in return. It was full of freeze-dried food, but it was quite a small box. Sarah eventually showed up, and when she saw it, she was a bit taken aback.

'Is that all there is?'

It looked as though someone had left a zero off the order. But she got in touch with Backcountry, who very generously agreed to pick up the tab if we went and got what we needed of their product from the Taupo branch of Hunting & Fishing. So that's what we did, and we just about cleaned them out.

'We'll need more containers,' I said, scratching my head as I looked at the pile of foil packets of dried meat and vegetables.

'Oh, another thing,' Sarah said. 'He's run out of water. His desalination plant isn't working properly.'

'Well, that's a bit different, Sarah,' I said. 'The food doesn't weigh much, but water ... How much does he need?'

'He reckons about 60 litres.'

'That sort of changes everything, Sarah.'

Scott had an ex-Armed Offenders Squad team leader, Nigel Escott, from Rotorua, acting as project manager. You could see why he was so successful in his former job, as he was very thorough and meticulous. He took over the coordination of the drop.

We packed the food and water in three containers. This time, mindful of how ill-advised our gift of chicken had been, we filled the voids with bread.

Mark and I flew out to the Taranaki coast on 20 June and then far out over the Tasman looking for Scott. We eventually found him, like Shaun had been, just a speck in the wide blue sea. We did three passes and supply drops. Some of the containers—water, as luck would have it—burst.

We made another drop of water to him six days later and yet another on 6 July. By now he was nearing the coast, but conditions turned against his attempt. He got within 60 kilometres of the Taranaki coast, and could actually see Mount Taranaki, but a strong nor'easterly set in and started driving him towards the top of the South Island. By Thursday, 10 July, we were sufficiently worried about him that I got up at 3am and began monitoring his progress and the deteriorating weather situation. He was having a hell of a time of it. The harness that kept him securely in position in his cabin had broken, and he kept getting rolled over and thrown around the interior. He'd received facial injuries, and was generally bruised and demoralised. It wasn't lost on me that the last person to have attempted to kayak the Tasman, Andrew McAuley, had been lost somewhere in 2007, about the same distance out from the Fiordland coast. His kayak was found, but there was no trace of the young Aussie himself.

Mindful of the importance of the mental aspect of this kind of enterprise, and remembering what a boost it had given Shaun Quincey when we'd shown up to offer a bit of moral support, we decided we'd fly out and see how he was getting on. He was about 83 kilometres offshore, so close, but so far. The weather was atrocious, with gusts of around 60 knots and 5-metre swells. We did a couple of wide circles and spoke to him on the marine radio. He sounded pretty close to the end of his tether, and it was no surprise when we heard while we were inbound that he had decided enough was enough and his situation was too dangerous. He had activated his emergency beacon, and was in touch with the Rescue Coordination Centre in Wellington. They dispatched the Taranaki rescue helicopter. We were on station over him and were able to relay his precise position to the helicopter, which dropped a man into the water and winched them both to safety.

He was gutted, needless to say, but everyone agreed he had made the right decision.

A week later, after he'd recovered from his ordeal, attention switched to trying to recover the kayak. I had never meant to get as deeply involved in the whole project as I was, but I really felt for both Scott and Sarah. They were keen to recover the little vessel, for both sentimental and practical reasons: there was a fair bit of expensive gear aboard, and the kayak would be useful as a prop when Scott took to speaking about his adventures.

There was still a signal from the beacon, so Scott and I flew to its source. We found the kayak lying awash and on its side in the sea about 20 nautical miles (37 kilometres) north-west of the base of Farewell Spit. Scott was delighted, and made preparations to go and retrieve it. The owners of a Takaka-based charter vessel, the *Maggie Rose,* offered to take him, and he set off at around 4pm on 15 July. We flew out to supervise.

To everyone's great disappointment, though, there was no trace of the kayak. The *Maggie Rose* retrieved the beacon, but it had somehow come adrift of the vessel. The kayak was nowhere to be found. About a month later, however, on 29 August 2014, a fellow helicopter pilot, Matt Newton, found it washed up on the Taranaki coast.

***

The skydiving company employee who had taken a swing at me when he failed our drugs test had gone backwards a bit since then. He hadn't worked since, and he was rumoured to still be smoking cannabis heavily. He was also embroiled, as they say, in a nasty custody dispute over a child he'd had with a friend of mine. She was in a new relationship and was seeking custody. She asked me if I would give evidence on her behalf in the family court in Rotorua. This was a tricky assignment, as it was a condition of our testing regime that the results would be confidential to us as the employer and to the person being tested. I couldn't just stand up there and say that he had failed a drugs test. But I was quite comfortable phrasing my answer in a way that conveyed the same information in code.

'Do you know the respondent, Mr Funnell?'

'Yes, he was an employee of ours in our skydiving operation.'

'And how did he come to leave your employment?'

'We introduced a drugs testing regime, and as a result, he no longer met the requirements to continue in employment with us.'

That was all his ex-partner's lawyer needed out of me. I wasn't quite so prepared for what his own counsel had up his sleeve.

'Mr Funnell, it's well known that you are a sexual deviant, isn't it,' he said.

I blinked.

'Well, Your Honour,' I said. 'I'm not sure how to answer that question. It's true that my marriage to my wife has ended and I am now in a new relationship. I've had the odd girlfriend in the past. I don't know whether that's sexual deviancy. I suppose it might be. But I'm just a poor, humble helicopter pilot going about my business.'

There was a murmur of laughter in the courtroom, and the judge gave me a sympathetic smile. When his written judgment in the matter was released, he even wrote that 'as a matter of record, the court finds that the allegations that Mr Funnell is a sexual deviant are totally unfounded'.

I'm not sure Trish Funnell would have agreed. I'm not proud to say that after four decades of marriage, Trish and I had separated by this time. It wasn't acrimonious, but it was final.

The end of our marriage, together with the death of our loyal in-house solicitor Hugh Barclay and the departure of our highly valued marketing manager Fritz, was also the end of what had been a very effective team. In the end, I'm good at putting teams together to carry out what I want done, but without the team, I'd be nothing. Not only had Trish been a brilliant mum to our boys, but she was a superb business partner. I tended to come up with what she called 'hare-brained schemes': she provided balance to my enthusiasm, and when she saw merit in any of my far-fetched ideas, she had a very clear view of the practical steps that would be required to realise them. I was quite good at making money, but she was even better at looking after it. She had a great eye for opportunities of her own. And, of course, she was a vital member of the search and rescue team. Right from the early days, whenever I managed to fly to someone's assistance, it was because Trish had been the person to receive the communication and alert me. Then, once the Philips Search and Rescue Trust was set up, she was for many years a very effective administrator. I got far more than my fair share of recognition for search and rescue activities: too often, the team aspect of it was overlooked. Barry Shepherd has been helping people for 30 years and has never been recognised. And apart from a joint commendation by the Order of St John in 1991, the essential part that Trish played has never been properly recognised, either.

Trish ended up with Jeff Winter, which I was pleased about. The relationship that precipitated our break-up didn't last long, but some time later I struck up with Chrissy

Simmen, who I had first met way back in 1983. She was working in the Skotel next to the Chateau at the time. She heard a helicopter coming in, and remarked on it to her workmate, Barb.

'Oh, that'll be John Funnell,' Barb said. 'He'll probably want a cup of tea.'

It was me. I landed, and Chrissy and Barb drove up to see me. Chrissy remembers grinning at the sight of me striding along as I have a tendency to do when I'm preoccupied, with poor old Dean Sparks hurrying along behind me trying to keep up.

Barb knew me a bit, and we all had a cup of coffee. Then Barb cadged a lift for herself and Chrissy back to town so that they could go shopping. It was Chrissy's first ride in a helicopter.

Newly married, Chrissy went overseas soon after that. But her marriage ended and she returned to Taupo. It's a small town and we had mutual friends. But we were properly reintroduced around 2006, and ended up moving in together. Chrissy reckons I told her that if she took up with me, it would be an exciting and interesting ride. I told her there would be challenges, but we'd mostly sort it all out. And I think it's probably worked out that way. For the best part of 10 years, she has been another vital member of the team.

***

In 2016, soon after I turned 65, I made the decision to step back from the aviation business altogether. It was up to Hamish and Mark now. Once again, it's a challenge being a mere observer in an outfit where you're used to being the boss. They don't do things the way I'd do them. Some things they'll do worse, but some they'll do better. The challenge for me has been to let go, and let them do it their way.

That's what it's all been about for me, really: the challenge. In the division of our assets I took over the hotel from Trish, and running this kind of enterprise is a challenge: Taupo is a bit of a tourist Mecca and we have a great location, but it's also a brutally competitive market. About seven years ago, I bought a farm just outside Taupo as a back-up plan, in case the Taupo Airport Authority went crazy with silly operational requirements and jeopardised our aviation interests. And in 2014, Chrissy and I got into bee-keeping as a hobby. The trouble with being the kind of person who needs to do things properly is that hobbies can get out of hand. We now find ourselves running an increasingly sophisticated honey extraction operation as well. The appeal of farming and bee-keeping and honey extraction is a challenge. Now that I've decided to do it, I'm trying to do it to the best of my ability.

The greatest challenge—and, therefore, the greatest satisfaction of all, when it went right—has been helping people. I still do a bit of it. Back in the days when I attended air accidents that pilots had survived, I would note the huge difference that there often was between the story the pilot told me and the 'official' story that came out once the investigators had been over it. I know first-hand how stressful it can be when you're the subject of official scrutiny and suspicion. So I have made a habit of offering support and advice to pilots who are under the gun, and I have also reached out to the families and loved ones who have lost a pilot in an air accident.

I learned to fly when flying was dangerous and sex was safe. How times have changed! I still fly. It's in my blood, and still a joy. I don't do much. I have flown well over 20,000 hours, but probably only fly 50 hours a year these days. I get myself from place to place in my little plane, but apart from that, I keep my feet pretty firmly on the ground.

Occasionally, people come up to me and ask me whether I remember them: they tell me I flew my helicopter and was instrumental in getting them out of some bad situation or another. I have an odd sort of memory, and I will usually draw a blank when I am bailed up like that. But once they start describing the details, it all comes back. I can remember the time of day, the weather, the technical difficulties associated. I can remember the sense of urgency, the adrenaline, the anxiety, the hours of boredom, the moments of terror. I can remember the faces, and I can remember the feelings. I remember every single one of them that we lost. And I remember every single one of them that we helped.

You don’t forget that stuff, in the end.

ACKNOWLEDGEMENTS

In writing the story of my life, thanks are due to the many volunteers who made possible the many rescues that are described, and the many that are not. It is bad form to single out one or two people, as there are many who stepped up to help when a helicopter retrieval was the only option—or in other cases the most expedient form of transport—to get the patient to the appropriate medical facility for their injuries. Nevertheless, Pat Wynne and Barry Shepherd deserve special mention as they have participated in many rescues with me. That is not to understate the work of dozens of others, ranging from cleaners, office staff, engineers, the New Zealand Fire Service and Police, St John New Zealand, to doctors and nurses, who had to put their trust in me when they flew with me, often in demanding conditions, to do my job and get them home safely.

I must also thank the many sponsors of our rescue activities, of whom only a few are mentioned in this book; the media, who played a big part in letting the public know what we were doing; and the trustees, most of whom gave their time free of charge.

Without the help and commitment of everyone named above, none of what is described in this book would have been possible.

Last of all, thanks to John McCrystal, who has done most of the work to bring this book to fruition.

One should never be afraid to stand up and be counted.

John Funnell

Taupo, 2016

BACK COVER MATERIAL

Often referred to as a 'search and rescue daredevil', John Funnell is one of New Zealand's longest serving and most respected search and rescue pilots, having clocked an incredible 19,000 hours of flying time.

In *Rescue Pilot* he shares stories from over four decades of flying search and rescue helicopters all over New Zealand—and beyond. John is a hero to the thousands of victims he has transported to safety over the years, and is perhaps best known for his unprecedented 800-kilometre mission to save a MetService employee attacked by a shark on the remote subantarctic Campbell Island.

Fearless and prepared to give pretty much any mission a go, John boasts a remarkable aviation career. What's more, he's a natural-born story-teller, and the adventures he recounts are utterly gripping.

www.ingramcontent.com/pod-product-compliance
Lightning Source LLC
LaVergne TN
LVHW080018210926
848481LV00011B/3051

* 9 7 8 1 5 2 5 2 3 0 0 1 1 *